D0908963

INTERNATIONAL ENVIRONMENTAL CONSULTING PRACTICE

INTERNATIONAL ENVIRONMENTAL CONSULTING PRACTICE

How and Where to Take Advantage of Global Opportunities

Peter A. Sam, ASc, BS, MS

JOHN WILEY & SONS, INC.

New York / Chichester / Weinheim / Brisbane / Singapore / Toronto

This book is printed on acid-free paper. ♾

Published simultaneously in Canada.

Library of Congress Cataloging-in-Publication Data:

Sam, Peter A.
International environmental consulting practice : how and where to take advantage of global opportunities / by Peter A. Sam.
p. cm.
"A practical text book for educators, career placement officials, environmental practitioners, environmental companies, manufacturers & firms."
Includes bibliographical references and index.
ISBN 0-471-17984-1 (cloth : alk. paper)
1. Environmental management. 2. Environmental impact consultants.
GE300.S36 1998
363.7′05--dc21 98-26716

Printed in the United States of America.

10 9 8 7 6 5 4 3 2 1

CONTENTS

PREFACE

As environmental management becomes an increasingly important element in forging each nation's economic, social, and human development efforts, the need for environmental experts increases, especially in nations where environmental technologies and experts are lacking. Nations generally classified as developing countries typically have extensive need for foreign environmental consultants. In fact, the demand for international environmental consultants has steadily risen in the past 10 years, owing to the complexity of environmental management in these countries. Such complexity arises from the integral role of socioeconomic development considerations in environmental management and a nation's overall sustainable development program.

It is understood that developing countries deal with unique socioeconomic issues—such as poverty, population explosion, and economic inefficiencies—that have an impact on natural resources. In developed nations, there are similar synergies between development and natural resources. However, in those countries there have been specific and rapid interventions over the years to institute stringent environmental policies and management strategies. In an earlier period of policy development in most of the industrialized countries, we witnessed a drastic increase in demand for environmental practitioners, especially in the United States. As the environmental field in some developed nations has become more or less saturated, we have started to see decreasing need for environmental consulting and related services in those industrialized countries.

This book provides the reader with comprehensive information on environmental opportunities in the international marketplace. It gives directions on how to set up an international environmental consulting practice and gives the reader an understanding of the evolution of global environmental management problems, issues, treaties, and agreements. The reader is presented with information on future opportunities within the international environmental consulting market. Analyses and trends that shape the present international environmental consulting market are provided to give the reader a broader view of the fields and future market certainties and uncertainties. A bibliography provides in-depth literature on consulting market opportunities as they reflect the complex socioeconomic dimensions of environmental management in various locations.

Chapter 1 consists of an introductory discussion on how environmental consulting practices facilitate global environmental management decisions.

Chapter 2 discusses the forces that shape and address global environmental issues. It explains the driving forces that trigger international environmental consulting and discusses the global environmental industry and markets. The role of the international donor community in creating consulting opportunities in the international market is presented, and the reader is introduced to the three major entities and institutions—international development assistance agencies; foundations, charitable, and other nonprofit organizations; and foreign governmental institutions—that create opportunities for consulting work within the global environmental market.

Chapter 3 provides the reader with detailed principles to consider before entering into an international environmental practice, together with methods of approach in preparing for international environmental consulting practice.

Chapter 4 discusses alternative business schemes to consider in entering the international environmental marketplace; included are international business partnering and joint ventures.

Chapter 5 deals with the basic principles of developing proposals and contracts for environmental projects. It provides the reader with techniques of bidding and negotiating in the international arena of environmental consulting practice.

Chapter 6 considers vital economic and market trends and provides an analysis of the global environmental market. It provides the reader with a broad, yet simplified, assessment of the international environmental consulting market up to the year 2000 and beyond, showing where the opportunities lie in the field. The assessment of market trends is sector-specific and focused on priority growth sectors.

Chapter 7 looks into the future of the international environmental market and discusses environmental segments that forecast promising opportunities as we enter the next century.

Chapter 8 provides the reader with useful information in preparing for a first business trip to a foreign country and an overview of the socioeconomic, environmental, and political orientation of each region of the world.

Pertinent environmental data and information for quick reference are provided in the Appendixes. These sections offer a variety of preliminary information the reader needs in order to understand the dynamics of the international environmental consulting arena and its "inner workings." They also provide information on international environmental laws, treaties and agreements, foreign ministries of environment and health, international environmental organizations, and other information relevant to the global environment.

The structure of this book is intended to make it easier for the reader and those interested in international environmental consulting to understand the field from various angles and to use the book as a reference manual. It offers a comprehensive approach for those wishing to enter the arena of international environmental consulting and environmental technologies exports and marketing. It also assists in the identification of those areas where environmental business opportunities exist within emerging markets, such as Africa and other developing countries.

This book provides specific information on the various aspects of environmental consulting for students with college degrees in environmental fields, college placement and career staff, environmental companies, environmental professors/lecturers, nonprofit environmental organizations, environmental product manufacturers, corporate environmental marketing representatives, and universities interested in environmental research. It also provides practicing environmental consultants with a guide to help in identifying opportunities in the international environmental marketplace. It discusses the evolution of international environmental consulting, and where and how financial resources can be obtained to establish a consulting practice in a foreign country. It also includes a list of U.S.-based national and international agencies that provide opportunities for international environmental projects.

ACKNOWLEDGMENTS

I give thanks and praise to God for the creation of the planet and the many blessings it brings to humankind. Praise also to the Almighty God for the wisdom he has provided environmental practitioners to protect the natural elements of the planet he created for the survival of his children and generations to follow. Special thanks to my parents—my mother, Martha, and my father, Peter A. Sam Sr.—for their most gracious support, guidance, and continuous advice throughout my life. Without them, this book would not have been possible. Thanks go to all friends and supporters. To all principal associates of the African Environmental Research Consortium (AERCG) (a nonprofit organization), those individual support and encouragement surpass my imagination—I salute all of you for the patriotic and voluntary efforts in environmental and human health protection in Africa. To all the peer reviewers and those whose names I have not mentioned, many thanks. My late sister, Elizabeth B. Sam, was pivotal in my early writings and publications. She was my adviser, strategist, and major critic. Her sudden and unfortunate death gave me the wisdom and empowerment to complete this book.

ABOUT THE AUTHOR

Peter A. Sam Jr., born in Ghana, West Africa, is a naturalized citizen of the United States of America. He holds a bachelor's degree in biology and chemistry from the Regis University and a master's degree in Environmental Planning/Sciences from the University of Colorado. Sam has worked in both private and public sectors for more than 13 years and has been in private consulting and international environmental practice for years. He is expert in urban and rural environmental issues. In 1993 he formed and organized the African Environmental Research Consulting Group (AERCG), a nonprofit organization comprising African expatriates and sympathizers of many nationalities. He chairs the organization and provides leadership and direction in advocating for fairness in international environmental treaties, agreements, and laws and in protecting human health and the environment in the developing world.

Sam has worked with international development assistance organizations and has interacted with international donor communities and multinational corporations. He has written several technical articles and publications in the field of environmental management, especially concerning developing countries, and has made many public presentations on international environmental management issues. He has wide experience in working with United States agencies and other countries on environmental management, environmental regulations, statutes, and standards (at the federal and state levels), and with nongovernment organizations (NGOs) and government agencies and departments worldwide. His portfolio includes direct consulting within the international arena. He is currently a senior environmental scientist with the United States Environmental Protection Agency (EPA) Region VII.

1

INTRODUCTION

The identification of international environmental issues began shortly after environmental awareness became a major concern in the Western hemisphere. In the late 1960s and early 1970s, the United States recognized the importance of environmental management. In 1970 it was apparent that environmental management and economic development were interlinked. After the United States commenced to address environmental management as part of socioeconomic development, this awareness began to spread worldwide. This global awareness led to the United Nations' quest to address the issues of global environmental degradation. As a result of the United Nations' Agenda for Change, most nations began to look at long-term strategies. For economic and political reasons, poorer and median nations were slow in making environmental protection a priority. The United Nations (UN) led the way to the creation of an environmental committee, the World Commission on Environment and Development (WCED). The United Nations General Assembly mandated the WCED to develop "Global Agenda for Change" to respond to global socioeconomic development and environmental issues. The Global Agenda for Change called for:

- Long-term environmental strategies for achieving sustainable development by the year 2000 and beyond
- Development of schemes for greater cooperation between developing countries and between countries at different stages of economic and social development
- Development of common and mutually supportive objectives among countries that take into consideration the interrelationships between people, resources, environment, and development

- Creation of opportunities and avenues for the international community to deal effectively with global environmental issues
- Development of a global vision and long-term shared perceptions for the protection and enhancement of the environment

In 1968 it became apparent that environmental degradation affected economic and social development. The importance of protecting of our natural resources for the next generation became an international issue. In 1968, a UN General Assembly resolution called for a Human Environmental Conference, which led to the 1972 Stockholm UN Conference on Human Environment. The conference resulted in a Declaration on the Human Environment, signed by 133 industrialized and developing nations. In addition to the Declaration, the United Nations Environmental Program (UNEP) was created, and so was the concept of sustainable development.

Subsequently, several international treaties and agreements have been signed. Most industrialized nations have enacted legislation and initiated environmental policies and regulations. Developing countries, however, have been slow to respond and to develop stringent environmental policies and regulations. In developing countries where environmental policies exist, there are few or no enforcement mechanisms to implement them. In many cases, the problem of weak environmental policy is due to a lack of expertise and weak governmental institutions. Even so, developing countries realized the need for environmental stewardship as it became clear that natural resource management is an integral part of social and economic development. The concept of building on positive synergies between socioeconomic development and the environment led multilateral financial institutions to expand their development activities toward environmental sensitivity and environmental benefits. The World Bank's specific intervention included environmental strategies in developing countries, especially in Africa, which led to the creation of the National Environmental Action Plans (NEAPs). The NEAPs have been the cornerstone of environmental capacity building in developing countries, especially in Africa. An NEAP considers key aspects of a country's environmental priorities and serves as a foundation for environmental policies and programs. The objectives of an NEAP are to provide a framework for integrating environmental stewardship into a country's socioeconomic development and to set priorities for an agenda that includes sustainable development. The next action plan for the developing countries in Africa is to build capacity in key environmental institutions. This process must rely on international environmental consultants, inasmuch as most of the developing countries lack local environmental expertise and technologies.

As developing countries develop their infrastructures and institute economic structural adjustment programs, the need for environmental management increases. There is also a greater need for input from foreign environmental consultants. International lending institutions and donor agencies have all developed environmental policies for financing projects. Most financed projects require an environmental impact assessment (EIA). Currently, the World Bank and other multilateral institutions are incorporating and addressing environmental issues in all their program implementation to ensure that potential adverse environmental impacts from their financed project activities are addressed up front. Other multilateral international lending institutions and donor agencies, such as the African Development Bank, have also created similar programs. Overall, environmental policies of multilateral financial institutions have caused developing countries to realize the importance of environmental protection and sustainable development. Many developing countries have cooperated with international environmental experts in their attempts to strengthen their internal institutional capacity and ultimately to improve the environment. It is anticipated that developing countries will continue to seek international consultants for environmental management, institutional capacity building, and environmental technologies.

1.1 THE NEED FOR INTERNATIONAL ENVIRONMENTAL CONSULTANTS

The environmental policy framework in the United States and many other developed nations—for example, Organization for Economic Cooperation and Development (OECD) nations—has brought about progress in environmental management in their respective countries. Developing countries are gradually moving toward the attainment of well-designed and aggressive environmental policies and institutions. The process is slow and difficult, in that many developing countries lack the necessary technical capacity and expertise. These countries therefore have relied, and will continue to depend, on foreign consultants to provide technical assistance and institutional capacity building in all sectors of environmental management.

It is apparent that the United States leads the world in environmental technologies and management experience, especially in the areas relating to the development of environmental regulations, requirements, standards, and legislation. But as the market becomes saturated with competing environmental technologies, practitioners, and companies in the United States, the U.S. environmental industry has begun to move into foreign markets, where there are growing opportunities for international environmental consulting practice. In fact, as the developing countries improve their infrastructure and integrate their environmental management strategies into the overall socioeconomic development programs, they are likely to provide major environmental consulting market opportunities.

The other major contributing factor to the surge in international environmental consulting opportunities is the intensification of global environmental awareness, fueled mainly by nongovernmental, proactive environmental groups worldwide. This movement has created an international political consciousness, which has resulted in the enactment of international laws, treaties, and agreements to tackle environmental issues. Global environmentalism has led to key global environmental activities that have increased the need for international environmental consultants. Among the relevant factors and activities that have generated this need are the following:

- *Global climate change:* A global initiative has been designed to establish cost-effective strategies by UN member nations to reduce greenhouse gas emissions.
- *Ozone depletion:* A global initiative was established under the Montreal Protocol of 1987, which sets schedules for reducing consumption of ozone-depleting substances.
- *Conservation of biodiversity:* A global initiative was established by the Earth Summit in 1992 and ratified by 157 countries to conserve biological diversity, protect and sustain ecosystems, and to ensure that benefits from the use of genetic resources are appropriately shared in the world community.
- *Desertification:* A global initiative was created to establish mechanisms for monitoring and assessing information dissemination, research and technology development, drought relief and education.
- *Agenda 21:* This is the global environmental agenda created at the Earth Summit of 1992, the latest meeting of the United Nations Conference on Environment and Development in Rio, Brazil. The objectives of Agenda 21 have created an immense opportunity for international environmental consulting services.

Other factors have been fueled by global economic activities as well:

1. The emergence of economic globalization, which is sweeping across industrialized countries such as those of the OECD community
2. The quest by developed nations (mostly the wealthy countries) to seek alternative competitive advantage in the expansion of their economies while maintaining a strong environmental stance
3. The increase in global economic activities, resulting in stringent international and local regulations because of global climate change/ozone treaties and agreements requiring air pollution control and management

What does this mean for environmental markets in developing nations? First and foremost, the international community, especially the United States and most developed nations (e.g., the OECD countries), have a vested interest in applying development aid packages to remedy the rest of the world's socioeconomic and environmental problems. The developing countries' reliance on outside financing for environmental management projects creates an avenue for wealthy countries, which control most of the official development assistance, and a multilateral financial institutions to encourage investment in environmental technologies in developing countries. The future growth for environmental technology exports and environmental consulting as part and parcel of development aid packages to developing countries is thus expected to increase.

Overall, environmental issues have become high-priority items for international and regional institutions. They want to encourage governments to implement policies and actions that will promote sustainable development and develop positive synergies between sound economic growth and sound environmental protection. Proactive environmental organizations have led the fight to put pressure on their own governments and within the international arena. The United Nations Organization (UNO) has taken a lead role in forging international relations between nations with the objective of developing global environmental policies. These policies and laws have forced governments to respond to the cyclic issues of socioeconomic and environmental deterioration in many parts of the world. International environmental policy developments have had a tremendous effect on shaping international environmental markets to such an extent that government actions have created the need for experts in environmental fields to assist in the implementation of global environmental accords. Other driving forces that led to global environmental policies and laws are shown in Figure 1.1.

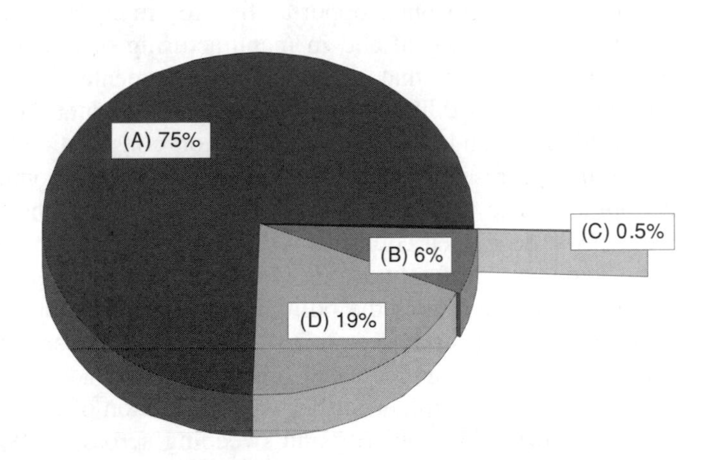

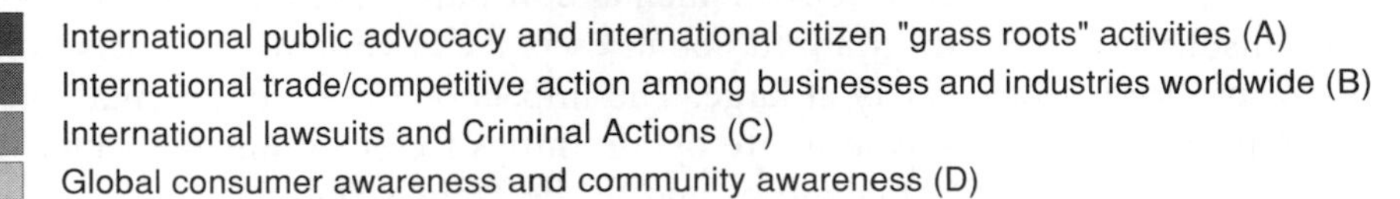

Figure 1.1 Driving forces of international environmental policy and laws (*Source:* P. A. Sam. African Environmental Research and Consulting Group (AERCG), 1996)

1.2 THE FUTURE FOR ENVIRONMENTAL CONSULTING PRACTICE

It is fair to say that the industry has come a long way from "pollute first and remedy later" to the "pollution prevention and waste minimization" approach. The mainstream environmental business within industrialized countries has, for decades, directed the fight against pollution through environmental laws, regulations, and remedial actions. Gone are the days when pollution remediation and anti-pollution regulations and laws accounted for the lion's share of environmental industry revenues. The paradigm shift has somehow led to an apparent reduction in demand for environmental consultants, especially in the United States, where hazardous waste and municipal waste (solid waste) consulting and engineering businesses have remained stagnated in many regions since 1994. The major environmental areas that apparently helped maintain market stability were U.S. government contracts—mainly for Department of Defense (DOD) and Department of Energy (DOE) remediation construction projects. The large environmental companies have survived, but the smaller ones have been affected tremendously by the decline in demand for environmental consulting services. Most smaller environmental companies gone out of business or have been forced to diversify or merge with other companies.

The slowdown in the environmental industry within some developed nations is expected to continue for years to come. Developing countries, however, are expected to open opportunities for many areas of the industry in research and development and in manufacturing of environmental technologies. It is noteworthy that while the environmental industry in developed countries shifts to pollution prevention, environmental consultation will remain in high demand, mainly in remediation construction, environmental energy sourcing, research and development, and environmental technology development. The most promising future markets for the environmental industry in the developed countries will be in waste minimization consulting.

The major reason for future growth in waste minimization consultation in the developed countries is the political and social pressures put on the manufacturing and industrial sectors to seek strategies for source reduction of waste within their manufacturing and production activities. These political and social pressures explain further why the notion of Life Cycle Assessment (LCA) is growing in popularity and sweeping across the Western world, providing a potential market for environmental consulting. Another area of hope for consulting in the industrialized countries is the environmental crisis born out of pressures from society at large. The impact on the environmental market is positive and will create future opportunities for environmental consulting practice within OECD countries and in the United States.

The developing countries, however, have began to take the traditional approach to establish regulations, build governmental institutional capacity,

Table 1.1 Categories of International Environmental Areas of Expertise

Categories	Areas of Expertise			
Earth Sciences and Mining Engineering	Survey Engineer Field Technician Economic Geologist	Geological Physicist Exploration Geologist Geoscience Administrator	Sedimentologist Soil Conservationist Coal Mining Engineer	Geotechnical Engineer Irrigation Water Manager Interpretive Geophysicist
Ecology, Forestry, and Wildlife Management	Forester Forest Engineer Animal Ecologist Aquatic Ecologist Plant Physiologist Forest Pathologist Wildlife Resources Technologist	Terrestrial Ecologist Conservation Volunteer Wildlife Preserve Manager Range Management Scientist Natural Resources Assistant Biological Control Specialist	Forest Products Technologist Botanist Zoologist Microbiologist Horticulturist Systems Ecologist	Estuarine Biologist Fisheries Scientist Developmental Biologist Environmental Biochemist Range Animal Nutritionist Quantitative Fish Ecologist
Energy Management and Engineering	Nuclear Engineer Nuclear Physicist Petroleum Engineer	Solar Energy Developer Uranium Mining Engineer Radiological Programs Engineer	Power Engineer Reservoir Engineer Geothermal Analyst	Petroleum Geologist Utility Corridor Planner Energy Conservation Economist
Environmental Engineering and Public Works	Sanitary Engineer Construction Engineer Environmental Engineer	Public Utility Manager Environmental Sanitarian Community Health Inspector	Land Surveyor Civil Engineer Solid Waste Manager	Public Works Director Architectural Engineer Water Resources Engineer
Marine Sciences and Engineering	Aquaculture Scientist Physical Oceanographer	Marine Systems Engineer Marine Surveyor	Marine Ecologist Chemical Oceanographer	

(continued on next page)

Table 1.1 (*Continued*)

Categories	Areas of Expertise			
Atmospheric Sciences	Climatologist	Air Pollution Control Analyst	Air Quality Monitoring	
	Air Quality Engineer	Meteorologist		
Environmental Planning and Administration	Landman	Environmental Lawyer	Recreation Planning	Coastal Zone Planner
	Park Ranger	Demographic Associate	Resources Policy Analyst	Agricultural Engineer
	Anthropologist	Environmental Designer	Architect	Environmental Economist
	Socioeconomist	Transportation Planner	Land Surveyor	Architectural Historian
	Science Educator	Environmental Lobbyist	Archaeologist	Interpretive Naturalist
	Rural Sociologist	Cultural Impact Analyst	Land Use Planner	Environmental Marketing Representative
	Population Analyst	Environmental Ombudsman	Photogrammertist	Water Resources Coordinator
	Landscape Architect	Environmental Geographer	Forestry Economist	Historic Preservation Officer
	Economic Development Specialist	Parks and Recreation Specialist		

Source: Peter A. Sam, 1996.

clean up/remediate contaminated sites, design and construct disposal sites such as hazardous waste and solid waste landfills, sewage treatment plants, and a multitude of other environmental projects. The lack of expertise in developing countries creates an opportunity for a future "boom" in international environmental consulting practice (see Table 1.1).

2

THE EVOLUTION OF INTERNATIONAL ENVIRONMENTAL CONSULTING

Since the early 1970s, international environmental consulting has increased in demand resulting, in part, from international treaties, agreements, and laws and a growing awareness worldwide that environmental pollution does not have borders and is therefore an issue for all nations. Furthermore, it became apparent that environmental management is of vital importance to national economic development and prosperity. In particular, environmental management and the preservation of natural resources are critical to the maintenance of human health. Consequently, international environmental consulting practices became important, in order that services could be provided to countries that do not have the local expertise that may be needed to deal with a particular problem in environmental management.

2.1 BACKGROUND TO GLOBAL ENVIRONMENTAL PROGRAMS

The end of the Second World War marked the beginning of reconstruction in socioeconomic and human welfare. Yet little attention was drawn to the environment and the conservation of natural resources. International and global environmental issues came to light in the 1960s, and subsequently the ideas of cooperation and technological development within member nations of the United Nations flourished. The Declaration by the United Nations on the Human Environment was formulated in 1972, and thereafter major global conferences were convened which led to the enactment of several other environmental laws, treaties, and agreements.

In the past 20 years, global environmental issues have gradually become part and parcel of international commerce and politics. These issues have become intricately linked to all other global socioeconomic and political issues, thereby increasing pressure on each government to maintain its commitment to addressing environmental issues within its political jurisdiction. The commitment by nations and governments to environmental preservation has led to the pursuance of sustainable development. The momentum of global environmental preservation has grown over the years and has contributed to institutional capacity building within nations to manage and control the negative impact of development and human activities. Many critical global problems have emerged, such as poverty, the resettlement of refugees, and the lack of human rights, all of which are related to development and the environment. The downward spiral in the quality of human life that accompanied the uncontrolled population explosion in urban centers worldwide has placed unprecedented pressure on the environment, resulting in environmental degradation, deforestation, and loss of biodiversity and other natural resources, especially in developing nations. Environmental issues were found to be linked to socioeconomic development; Figure 2.1 illustrates the relationship between economic activity and the environment.

Several years of examination and reexamination of the critical issues that lay between the environment and development led to the creation of the UN Global Environmental Agenda (Agenda 21), which offers a new approach to

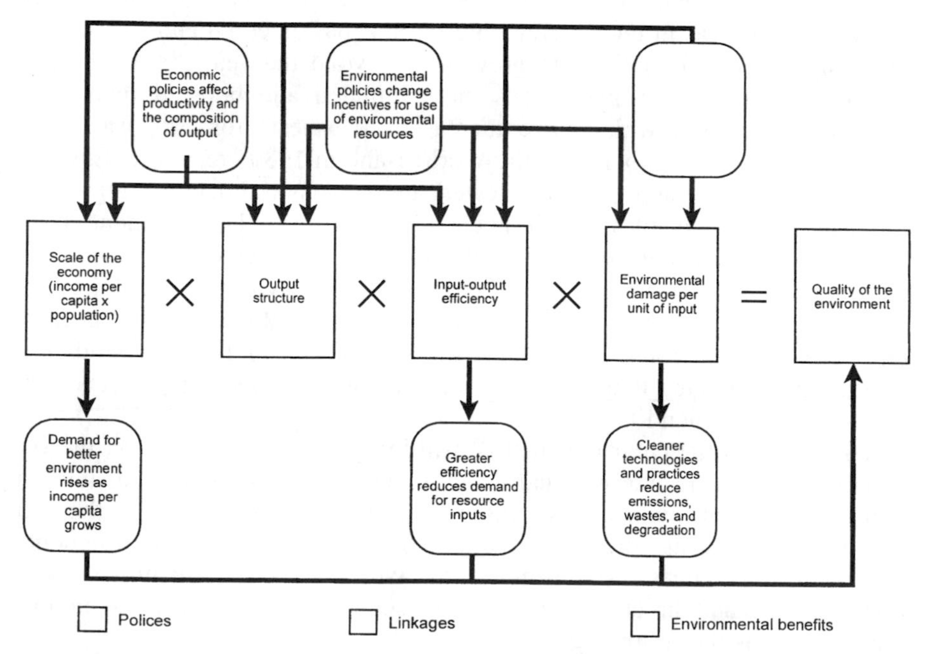

Figure 2.1 Economic activity and the environment (*Source:* World Bank staff)

solving global environmental problems. The new ways of thinking represent a more holistic approach, which embodies interlocking factors such as economic development, trade, poverty, and population and natural resource management, and are the basis of the present era of international environmental consulting practice in developing countries where the interlocking factors usually are ill-defined. This new era and new way of thinking have opened avenues for environmental practitioners to work abroad in regions of the world where expertise is lacking. The holistic approach is underscored by the sustainable development principles in Agenda 21 and relates to addressing environmental problems in countries where environmental problems can be attributed to unique social dimensions and economic stagnation. The crisis is highly felt in developing countries, such as those in Africa, that have experienced a high degree of economic stagnation (especially in the 1970s and 1980s), droughts, famine, desertification, degradation of soil fertility, low agricultural productivity, deterioration of coastal and water resources, and massive human suffering. The aggregation of the socioeconomic factors, among others, led to a massive environmental deterioration. It has became obvious that the solutions to the environmental and development problems must be approached concurrently.

As societies worldwide become aware of the links between economic development, human survival, and social development, and between poverty and environmental deterioration, the need for the inclusion of environment management and sustainable development in economic development programs has become a priority within many nations. The need for expertise and consultancy in these areas in most developing nations has emerged and is projected by many experts to continue to grow well beyond the year 2000. Many of the international donor institutions have initiated and boosted their environmental activities, which have increased "Requests for Proposals" and consultancy work. For example, the World Bank in 1980 created its Environmental Department and in 1993 created another environmental wing, Environmentally Sustainable Development. The African Development Bank followed in the footsteps of the World Bank, among other international institutions, to create the African Development Bank's Environmental Policy Paper. This action provided an opening for environmental consultancy, because the Bank used environmental impact assessment and preliminary studies to ensure and guarantee that economic and capital development projects were environmentally sound.

The international environmental consulting practice is steadily on the rise as governments and international development agencies—including international multilateral financial institutions—have commenced to tailor the economic development programs they support to sound environmental management practices. For example, the World Bank has instituted operational policies and directives relating to its socioeconomic environmental programs, such as the following:

- *Environmental Assessment.* This directive outlines World Bank policy and procedures for the environmental assessment and analysis of World Bank lending and operations. Environmental consequences should be recognized early in the project cycle and taken into account in project selection, sitting, planning, and design.
- *Environmental Action Plan.* This directive outlines World Bank policy and procedures for preparing national environmental action plans by borrowing governments. It is bank policy to foster the preparation and implementation of an appropriate plan in each country, reflect the findings and strategies of the country's plan in World Bank's, and provide technical assistance as requested. The plan aims to identify key environmental problems, set priorities for dealing with them, and identify appropriate investments.
- *Agricultural Pest Management.* The bank's policy promotes effective and environmentally sound pest management practices and advocates the use of integrated pest management techniques in World Bank's-supported agricultural development projects.
- *Involuntary Settlements.* This directive describes bank policy and procedures for involuntary resettlement, as well as the conditions that borrowers are expected to meet in operations involving involuntary resettlement. Where large-scale population displacement is unavoidable, a resettlement plan, timetable, and budget are required. The policy aims to ensure that the population displaced by a project is provided with the means to improve, or at least restore, their former living standards, earning capacity, and productivity levels.
- *Forestry.* This policy statement provides guidance to bank staff involved in forestry projects, detailing that the bank will not finance commercial logging operations or acquisition of equipment for use in primary moist tropical forests; in forests of high ecological value it will finance only preservation and light, nonextractive use of forest resources. The objectives are to provide for a sustainable stream of direct or indirect benefits to alleviate poverty and to enhance community income and environmental protection.
- *Procedures for Investment Operations Under the Global Facility.* This directive describes additional steps in standard bank investment lending procedures, including environmental assessment required to process Global Environmental Facility (GEF) operations.
- *Environmental Policy for Dam and Reservoir Projects.* This establishes policy for dam and reservoir projects and codifies best practices, including preparation of preliminary reconnaissance to identify potential environmental effects and ascertain the extent of needed environmental studies and actions. The policy requires creation of an independent panel of environmental experts for large dams and other projects with major

environmental implications. Adverse environmental impacts should be avoided, minimized, or compensated for, wherever possible, during project design (such as by modifying dam location or height) and by measures implemented as part of the project.

- *Institutional Development Fund (IDF)*. This directive describes the purpose and use of the fund as a small grant facility for financing technical assistance for institutional development work in policy reform, country management of technical assistance, and areas of special operational emphasis, such as the environment.
- *Country Economic and Sector Work.* This work analyzes the macroeconomic and sector development problems of borrower countries. As the long-term quality and sustainability of development depend on factors in addition to economic ones, country economic work may also focus on questions of the environmental effects of alternative policy options.
- *Poverty Reduction.* This directive summarizes policy and guidelines for operational work in poverty reduction. Attention is given to the impact of sector policies on the links between environmental issues and poverty.
- *Technical Assistance.* Bank policy and procedures for technical assistance include preparation and implementation support for environmental action plans, various phases of the project cycle, and environmental assessment. Institutional development assistance may address the need to strengthen capacity for environmental analysis and policy enforcement.
- *Adjustment Lending Policy.* Analysis of adjustment programs considers implications for the environment. Bank staff are to review the environmental policies and practices in the country, take into account the findings and recommendations of such reviews in adjustment program design, and identify the linkages between the various reforms in the adjustment program and the environment.
- *Proceeding of Investment Lending.* This directive summarizes bank procedures and documentation required for investments and loans, including those for projects with environmental objectives and components, and provides guidance to bank staff. Procedures include determination of project environmental category, type and timing of an environmental assessment, and environmental issues to be examined. These are followed by a discussion of a project's environmental impact, the main findings of the environmental assessment, consultation with affected groups, and feedback to these groups on the findings of the assessment.

Other international development agencies have adopted similar approaches. This shift in thinking to a holistic approach to solving environmental problems has become a tool for achieving successful economic growth and human survival. The emergence of a new era of socioeconomic development and environmental management by international lending institutions, international development agencies, and other donors has led to operational policies that demand environmental considerations be made part of the provisions of any development loan or development assistance to recipient countries. The new lending practices that include environmental considerations have resulted in a growing trend and increased demand for international environmental consulting practice.

Overall, the climate for environmental consulting is likely to become even more favorable in developing nations, because they lag in environmental management resources and capacity and have further been placed in situations that warrant foreign environmental experts to assist and implement environmental plans and projects. For example, the World Bank National Environmental Action Plans (NEAP) initiative in Africa has led to increased demand for consultants to assist countries in planning and facilitating the development of NEAP. Preparation of national environmental strategies such as the NEAP is just a minuscule factor in dictating the extent and performance of the international environmental market. The next section of this chapter examines the major driving forces that shape and stimulate the international environmental market.

2.2 FORCES SHAPING INTERNATIONAL ENVIRONMENTAL CONSULTING PRACTICE

The Global Agenda 21, established in 1992 at the Earth Summit in Rio de Janeiro, Brazil, became the single most comprehensive document to address several global environmental issues. It provides a shift in the approach to global environmental problems. The document of more than 700 pages, approved by 180 nations, provided solutions and an agenda for global protection of the environment. The following are the major objectives of Agenda 21:

1. To propose long-term environmental strategies for achieving sustainable development by the year 2000 and beyond
2. To recommend avenues for addressing global environmental issues
3. To seek greater cooperation among developing nations and between nations with varying environmental management capacities and varying stages of socioeconomic development, with the optimum aim of achieving common and/or mutually supportive objectives that amount to interrelationships between human beings, natural resources, environment, and development at large
4. To consider avenues by which the international community can more effectively tackle environmental issues
5. To assist in defining shared perceptions of long-term environmental problems and issues and the appropriate efforts needed to address the problems of protecting human health and the environment

The principles of Agenda 21 encourage sustainable development. These are broad implementation principles, which include safeguarding the global ecology and providing directions for international policies and actions for needed change. The Agenda 21 principles have indeed led to a high demand for environmental consultants as governments develop policies and strategies toward sustainable development.

As the Global Agenda for Change (Agenda 21) gains momentum and environmental concerns become increasingly profound within the global society, the role of the prominent forces that stimulate worldwide environmentalism will increase. The most notable enviro-economic trends will include:

- The global integration of economic and trade policies with environmental provisions
- Sustainable development
- Waste-to-energy and pollution prevention
- Continued activities in waste minimization of final consumer products
- Increase in technology transfer, advance waste minimization, and Pollution Prevention technologies to all manufacturing and industrial sectors
- Increased technologies and consideration of Life Cycle Assessment in all manufacturing sectors
- Increased growth and faster gains in environmental technology exports to less developed countries (LDCs)
- Increased growth and faster gains in environmental technology manufacturing in developed countries such as the United States, Japan, and Germany
- Impact of Structural Adjustment Programs (SAPs) in developing countries as they shift production from goods to services

Ultimately, there are four prominent forces shaping international environmental consulting practice:

1. International Policies and Agreement (IPA)
2. Governmental and Non-governmental Roles (GR)
3. International Organizations (IO)
4. International Donor Community (IDC)

These forces are stimulating factors that promote activities under the broad umbrella of environment, economic, and human development.

What activities of these prominent forces shape international environmental consulting practice? It is important to understand the interrelationship between these forces. Overall, the forces that stimulate international environmental consulting practice operate in many ways to meet their objectives and goals, influencing global enviro-economics activities. Figure 2.2 illustrates the building blocks and structures of these forces.

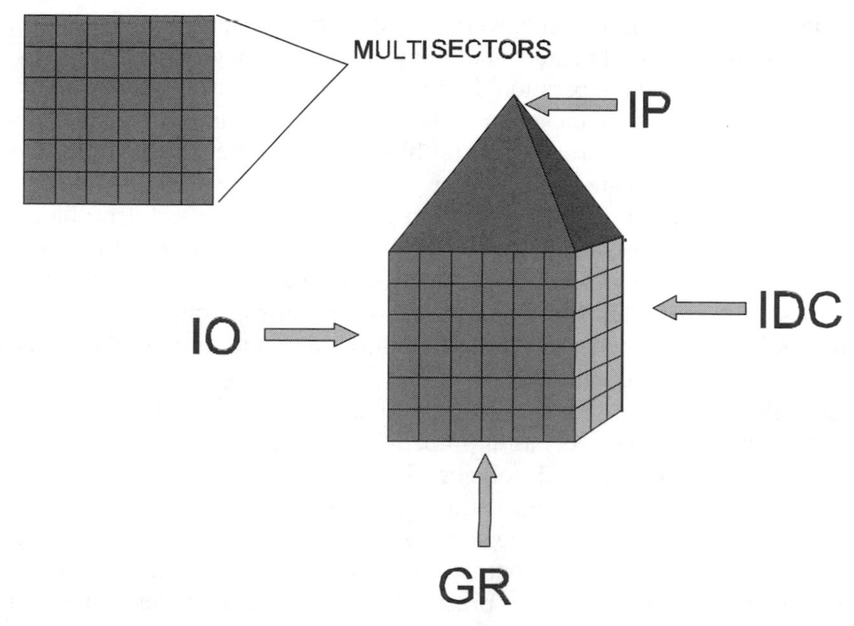

Figure 2.2 Forces creating markets for international environmental consulting

There are indeed many sectors which, when combined, provide a tremendous force for environmental market growth. The multisectors are the environmental activities within specific environmental media that create the market for environmental consulting demand. These activities are described in the following sections.

2.2.1 International Policies, Agreements, and Laws

The global environmental issues that confront the world have traditionally been addressed through international agreements, policies, and laws among nations. As of today, there are approximately 135 international treaties, agreements, and laws pertaining to the environment. International environmental laws and treaties between nations are usually implemented by each government of a country or nation, and their implementation creates markets for environmental management and consulting.

Since the 1970s, environmental concerns have been emphasized in international trade agreements and treaties, but universal standards for the protection of human health and environmental safety have not been adopted. In most instances, developed countries have the resources and capacity in terms of environmental technology, funding, and expertise, whereas in the less developed countries there is a scarcity of resources and overall capacity to establish stringent in-country policies, laws or requirements, and standards. Therefore, the market forces in developed and less developed countries tend to be rather different.

The major way in which governments, institutions, and individuals approach global environmental and socioeconomic development issues is through international cooperation. International policies, agreements, and laws can be characterized as the main standard protocols, that reflect the global approach to solving environmental problems. Governments, institutions, and other entities have created avenues for global environmental consulting practice through monitoring and research to identify environmental problems at local levels and define local situations in scientific and technical terms. Scientific environmental innovations and the development of new control technologies, processes, and products in many countries, both developing and industrialized, have been created through these laws and through the activities of the aforementioned entities.

Overall, cooperation among sovereign nations to seek solutions to global environmental problems acts as an impetus for creating international environmental agreements, laws and policies. Special Appendices Section XIX presents information on sovereign nations' cooperation and participation in key international conventions and regional agreements aimed at global environmental issues. International cooperation on environmental issues continues to grow and, in growing, stimulates practical environmental actions and activities that shape international environmental consulting practice.

The United Nations has been the driving engine in the establishment of international environmental laws through its United Nations Environmental Program (UNEP), whose leadership in advocating and coordinating environmental activities within the different branches of the UN and among nations has led to several international environmental laws, treaties, and agreements. Table 2.1 lists some of the major international agreements and laws that have been instituted directly or indirectly through the actions and leadership of the United Nations and its subsidiaries.

The mandates of these established international environmental agreements, laws, and policies also trigger a reaction by international task force groups and committees commissioned to develop a preamble to the rule-making and promulgation of the law. For example, the United Nation's first Conference on the Human Environment in 1972 provided an environmental management framework for the international community similar to that of the United States' National Environmental Policy Act of 1969, which, among other principles, stated:

> The protection and improvement of the human environment is a major issue which affects the well-being of peoples and economic development throughout the world.

The World Commission on Environment and Development (WCED) was created as a result of the United Nation's General Assembly Resolution 38/161 adopted in 1983. WCED developed legal principles for environmental protection and sustainable development that require nations, governments, and

Table 2.1 Selected International Environmental Treaties, Agreements, and Laws

Media	Environmental Law	Purpose
Human Ecology	Declaration of the UN Conference on the Human Environment (6/16/72, Stockholm)	Preservation and enhancement of human environment
Natural Resources	UNEP Principles of Conduct in Field of the Environment for Guidance of Nations in Conservation and Utilization of Natural Resources Shared by Two or More Nations (5/19/78, Nairobi)	Conservation and utilization of shared natural resources
Natural Resources	Nordic Convention (2/19/74, Stockholm)	Protection of the environment cooperation to prevent environmental damage by one nation to another nation
Human Ecology	UNEP Principles of Environmental Impact Assessment (6/6/87, Nairobi)	Promotion of procedures for EIA, including transboundary effects
Natural Resources	International Convention for Protection for Birds (10/18/50)	Protection of birds
Natural Resources	Convention for Conservation and Preservation of Soil, Water and Plant Life	Joint action by African nations for conservation and preservation of soil, water, and plant life
Natural Resources	Convention for Conservation and Preservation of Soil, Water and Plant Life	Protection of cultural and Natural Heritage
Natural Resources	Conference on Endangered Species	Protection of endangered species
Natural Resources	International Conference on Migratory Species	Protection of wild animals' migration across national boundaries
Natural Resources	International Conference on Cooperation among Nations of Timber Produce	Cooperation between timber producing nations
Natural Resources	International Treaty on Preservation of Plant Genetics	Preservation of plant genetic resources
Air Resources	Vienna Convention on Ozone Layer Protection (3/22/85, Vienna) and the Montreal Accord (1987, Montreal)	Control of global ozone depletion
Water/Marine	International Convention on Intervention on the High Seas in Oil Spills/Pollution and Casualties (11/29/69, Brussels)	Contingency plan action against maritime casualties resulting from oil pollution

(*continued on next page*)

Table 2.1 (*Continued*)

Media	Environmental Law	Purpose
Hazardous Waste	Basel Convention	Control of transboundary movement of hazardous waste and disposal
Hazardous Waste	Bamako Convention (1/30/91, Bamako)	Ban on import of hazardous waste to Africa and movement of hazardous waste within Africa
Nuclear Waste	IAEA Convention on Nuclear Accident (9/26/86, Vienna)	Provide early warning in nuclear accidents
Workers Protection, Health and Safety	ILO Convention for the Protection of Workers Against Ionizing Radiation (6/22/60)	Occupational safety and health
Workers Protection, Health and Safety	ILO Convention Against Hazards of Poisoning from Benzene (6/23/71)	Occupational safety and health
Water Resources	Agreement on the Action Plan for Environmentally Sound Management of the Common Zambezi River System (5/28/87, Harare)	Cooperation on environmentally sound water resources management of the common Zambezi River system for sustainable development
Water Resources	Convention on the River Gambia (6/30/78, Kaolack)	To ensure coordination and development of the Gambia River for sustainable natural resources
Water Resources	Agreement for the Establishment of an Organization to Manage and Develop the Kagera River Basin (8/24/77, Rusumo)	Establishment of the Kagera River Basin Authority for Cooperation Between riparian states
Water Resources	UN Action Plan on Water (3/25/77, Mar Del Plata)	Sustainable development of water resources and management
Water Resources	Convention on the Establishment of Senegal River Development Organization (3/11/72, Novakchott)	Economic development/sustainable water resource utilization
Water Resources	Agreement on Navigation, Economic Cooperation Between the States of the Niger (10/26/63, Niamey)	Cooperation for sustainable development and navigation
Aquatic Life	Convention for Conservation of Salmon in North Atlantic Ocean (3/2/82, Reykjavik)	Conservation, restoration, and management of salmon stock

Table 2.1 *(Continued)*

Media	Environmental Law	Purpose
Aquatic Life	Convention on Future Multilateral Cooperation in North/East Atlantic fisheries (11/18/80, London)	Conservation and optimal utilization of the fishery resources of North/East Atlantic
Aquatic Resources	International Convention for the Conservation of Atlantic Tuna (Rio de Janeiro, 5/14/66)	To protect and maintain population of tuna and tuna-like fish in the Atlantic
Marine/Water Resources	Convention for Marine, Coastal Environment of Eastern African Region (6/21/85, Nairobi)	Management/protection of marine environment and coastal areas in East Africa
Water Resources	Agreement for Cooperation in Dealing with pollution of the North Sea by Oil and Other Harmful Substances (9/13/83, Bonn)	Control of discharges of oil and harmful substances in the North Sea
Water/Marine Hazardous Waste	Convention on Marine Pollution via Dumping of Wastes (12/29/72, London)	Control pollution by marine/sea dumping
Hazardous Waste/Water/Marine	International Convention for Prevention of Pollution from Ships (11/2/73, London)	Marine environmental management against pollution
Hazardous Substance	Agreement for Transportation of Dangerous Goods (4/26/57, Geneva)	Provision to regulate modes of transportation
Hazardous Substance	Convention for Mutual Recognition of Inspections in Respect to Pharmaceutical Products (10/8/70, Geneva)	Removal of obstacles to international trade and quality control of pharmaceutical products
Hazardous Substance	FAO International Principles on the Distribution and Use of Pesticides (11/28/85, Rome)	Framework for control of pesticides
Hazardous Substance	UNEP London Guidelines for Exchange of Information on Chemicals in International Trade (6/17/87, Nairobi)	
RIO Convention on Environment and Human Development	UNCED, Agenda 21 (Rio de Janiero, 1992)	Conservation of biodiversity, climate change, desertification, conservation, and sustainable development

Source: World Resource Institute (1995 World Development Report).

international institutions to undertake environmental activities. The General Principles on environmental law adopted by the WCED are as follows:

1. *Fundamental Human Rights.* All human beings have the fundamental right to an environment adequate for their health and well-being.
2. *Intergenerational Equity.* Nations shall conserve and use the environment and natural resources for the benefit of present and future generations.
3. *Conservation and Sustainable Use.* Nations shall maintain ecosystems and ecological processes essential for the functioning of the biosphere, shall preserve biological diversity, and shall observe the principle of optimum sustainable yield in the use of living natural resources and ecosystems.
4. *Environmental Standards and Monitoring.* Nations shall establish adequate environmental protection standards and monitor changes in, and publish relevant data on, environmental quality and resource use.
5. *Prior Environmental Assessments.* Nations shall make or require prior environmental assessments of proposed activities that may significantly affect the environment or use of natural resources.
6. *Prior Notification, Access, and Due Process.* Nations shall inform, in a timely manner, all persons likely to be significantly affected by a planned activity and to grant them equal access and due process in administrative and judicial proceedings.
7. *Sustainable Development and Assistance.* Nations shall ensure that conservation is treated as an integral part of the planning and implementation of development activities and provide assistance to other nations, especially to developing countries, in support of environmental protection and sustainable development.
8. *General Obligation to Cooperate.* Nations shall cooperate in good faith with other nations in implementing the preceding rights and obligations.
9. *Reasonable and Equitable Use.* Nations shall use transboundary natural resources in a reasonable and equitable manner.
10. *Prevention and Abatement.* Nations shall prevent or abate any transboundary environmental interference which could cause or which causes significant harm.
11. *Strict Liability.* Nations shall take all reasonable precautionary measures to limit the risk when carrying out or permitting certain dangerous but beneficial activities, and shall ensure that compensation is provided should substantial transboundary harm occur even when the activities were not known to be harmful at the time they were undertaken.
12. *Prior Agreements When Prevention Costs Greatly Exceed Harm.* A nation shall enter into negotiations with an affected nation on the eq-

uitable conditions under which the activity could be carried out, when planning to carry out or permit activities causing transboundary harm that is substantial but far less than the cost of prevention.

13. *Nondiscrimination.* Nations shall apply as a minimum at least the same standards for environmental conduct and impacts regarding transboundary natural resources and environmental interferences as are applied domestically.
14. *General Obligation to Cooperate on Transboundary Environment Problems.* Nations shall cooperate in good faith with other Nations to achieve optimal use of transboundary environmental interferences.

The policies, laws, and agreements serve as a demand-driven machine for international environmental consultants. This machinery, which dictates the international environmental market, is largely felt in developing countries. Without exception, industrialized countries are also affected to some extent. For example, industrialized countries develop foreign trade policies and marketplace strategies that tend to provide them with a competitive advantage in the world market. In some respects, the same thinking becomes part of formulating development aid programs to address the socioeconomic and environmental issues of less developed countries. The environmental market is one of the several markets emerging rapidly as a result of international trade policy, economic policy, environmental policy, and social policy. Implementation of these policies in each country generates markets for other areas of the international environmental movement, such as environmental technology exports, environmental technologies transfer, environmental management capacity development, and remediation/restoration.

The United Nations Conference on Environment and Development (UNCED) in Rio and Agenda 21 also stress trade and socioeconomic development, in addition to natural resource conservation, management, and development. International trade agreements that contain environmental provisions result in opportunities for the environmental market and consultancy. These opportunities increasingly include international environmental consulting in areas of environmental law and general environmental management for experts and consultants who have expertise in assessing, for example, the trading in goods and exports that may pose a threat to human health and the environment. Most typical of these exports and trading goods/products are pesticides, hazardous chemicals, and other agricultural products. The commitment to provide financial resources, technology transfer, and related support to the less developed countries is a force that will drive international environmental markets in the future.

Ultimately, the international environmental consulting market, unlike other markets that are driven by demand, is driven by international policies, laws, and treaties. These forces trigger other nations' governmental policies and programs. These are the foremost and basic impetuses that create a demand for environmental consulting activities. There are other forces that also shape the demand for the environmental consulting market in both developed and developing countries, such as citizens advocating and pressuring governments to set environmental protection as a high priority on the national agenda and seeking protective regulations and standards for industries. This activity, in most cases, occurs in developed countries. In developing countries, the pressures are more likely to originate from international multilateral financial institutions and from the international donor community.

In general, global environmental issues and concerns are addressed through international agreements and policies by a number of different international organizations. As policies are formulated, agreements are signed between nations. The implementation of these agreements triggers various activities and environmental market forces, such as in-country environmental policy development, establishment of rules and regulations, and setting of environmental standards, and creates the market for environmental consulting. In developing countries where resources and expertise are lacking, there is a greater market for foreign environmental experts, which thus creates a market niche for international environmental consulting.

2.2.2 The Role of International Organizations

Since the 1970s, international organizations have had specific roles in addressing global environmental issues, such as in the development of programs and activities that inspire and encourage in-country governments to develop policies and embark on promoting sound environmental management and sustainable development. Through their mission, these organizations create numerous opportunities for international consulting—they drive the international environmental market. Most of these organizations seek consultants in developing their programs and various environmental management activities, which require environmental services in areas such as the following:

- Preliminary assessments
- Feasibility studies
- Program design
- Research
- Monitoring
- Evaluation
- Fiscal implementation

Typically, several environmental sectors are addressed, including the following:

- Land degradation
- Water resources management
- Hazardous waste and toxic chemicals
- Oceans and coastal waters and zones management

- Air pollution management, ozone depletion, climate change, transboundary air pollution, and acid rain
- Biodiversity
- Sustainable development
- Pesticides management
- Deforestation

It is interesting to note that the role of international organizations has become more important after the collapse of the Soviet Union and the end of the Cold War. There are many reasons for this shift. For instance, evidence that pollution does not have boundaries and the global recognition of environmental deterioration, poverty, and loss of biodiversity in many geographic regions of the world have strengthened the role and importance of international organizations. International organization functions and objectives are sometimes complex, but the thrust of their objectives for the environment, to a large extent, includes:

- Encouraging national governments to pursue sustainable development and develop sound policies for the management of the environment
- Providing the mechanism and leadership in integrating environmental concerns with economic and social development
- Promoting cooperation between nations on environmental and transboundary issues
- Providing resources and technical assistance for in-country institutional capacity building
- Coordinating and developing policies on issues affecting the environment and socioeconomic development.

The international organization is in part led by the United Nations, which comprises major distinct functional units and departments and/or agencies that have specific and specialized tasks. The specialized task departments and agencies undertake environmental activities such as management of natural resources, biodiversity, and climate change, and programs geared toward protection of human health and the environment.

Some of the environmental activities of select international organizations that play significant roles in addressing global environmental issues are shown in Table 2.2.

2.2.3 The Role of Governmental and Nongovernmental Institutions

Nations that are members of the United Nations maintain their sovereignty and autonomy. Therefore, apart from participation in international conferences on the environment and international environmental policies, national governments have the ultimate responsibility to protect human health and the environment within their own countries. Responsibility for natural resources and

Table 2.2 Examples of United Nations and Other Intergovernmental Bodies

Agencies	Programs
Economic Commission for Europe	Air and water pollution control; urban development
International Maritime Development of International Organization (IMCO)	Information on oil spill control; development of international controls over marine pollution
UN Educational, Scientific, and Cultural Organization (UNESCO)	Scientific studies on changes in oceans; development of man and the biosphere
UN Development Program (UNDP)	Resource surveys, conservation projects, ecological studies, training programs
World Health Organization (WHO)	Definition of environmental standards for human health; identification of environmental hazards—air, water, soil, and food pollution, such as caused by pesticides; studies of effects of induced changes in the environment, such as rapid population changes, massive migration, and industrialization
Food and Agriculture Organization (FAO)	Conservation soil, forests, and terrestrial waters; studies on water quality criteria for fish, pulp and paper mill effluents, and sewage effluents
UN Conference on the Human Environment	Preparatory work for the 1972 Stockholm Conference
World Meterological Organization (WMO)	World Weather Watch information related to air pollution and maritime environment
NATO Committee on Challenges of Society	Disaster assistance, air pollution, open, modern society water pollution, inland pollution; Regional Environment and Development Strategy
Organization for Economic Cooperation and Development (OECD)	Water resources management, air pollution, pesticides; organizing an Environmental Division
Organization of African Unity (OAU)	Serves as the depository for the African Convention on the Conservation of Nature and Natural Resources
Organization of American States (OAS)	Latin American Convention on Nature Protection and Wildlife Preservation in the Western Hemisphere
Council of Europe	Information sharing, pollution control, ecology
The European Community	European economic and social cooperation

Note: Refer to Appendix XIX for additional data.
Source: United Nations (U.N.), UN Chronicle (U.N., New York, June 1993) and UN Economic and Social Council, Programmes and Resources of UN System for the Biennium 1992–1993: Report of the Administrative Committee on Coord (U.N. NY, 1993, pg. 6).

socioeconomic development rests with national governments. The dynamics of environmental issues differ between developed nations and developing nations. Developing countries are far behind in environmental pollution prevention and natural resources management. Environmental policies in many developing countries are in the preliminary stages of development. Even in countries where environmental policies have been instituted, there is weak institutional capacity to provide policing and enforcement. Many industrialized countries, however, have developed environmental policies and instituted stringent regulations and standards. The general environmental situations facing developing countries differ drastically from those of industrialized nations; therefore, a universal policy would, in many instances, not be applicable. The question is, what role do national governments play and how do they manifest themselves in creating markets for environmental consulting?

Although developing countries have for many years centered their national priorities on economic development, the realization that economic development is interwoven with sound environmental and natural resource management has reached the hearts of national governments in these countries, so the challenges and environmental priorities have risen to almost the same level as in industrialized nations. Their role in environmental and natural resource management, infrastructure service provision, and overall sustainable development has become a function of governments worldwide. This role is an ongoing process that calls for a holistic approach to development. Environmental management's becoming a larger and growing part of the solution has brought about increased demand for environmental consultants and practitioners. National governments, both in the developed and industrialized countries, continue to seek experts within the environmental arena for consultation and implementation of environmental activities. The international environmental agreements that a country enters into or ratifies must be implemented in accordance with the international principles and the convention for which the particular international environmental law provides. In this process the national government frames and designs its agenda and initiates policy, rules, standards, and programs to meet the provisions of the law. Many national governments, especially in developing countries where technology development capacity is marginal, environmental experts are scarce, and environmental technologies are lacking, rely heavily on foreign experts and importation of environmental technologies, thereby creating in-country need for international consulting. As a result, such developing countries seek international environmental experts to assist in the development of their environmental activities. Industrialized countries also seek international environmental experts, but not to the same extent as developing countries. In fact, several international policies, laws, and agreements affect national governments' development programs, which in turn create the market for international environmental consulting.

Nongovernmental organizations (NGOs), private businesses, citizen action groups, and other organizations such as African Ministrial Conference On the Environment (AMCEN), the North Atlantic Treaty Organization (NATO), and the Organization of African Unity (OAU) also seek environmental consultants in different ways. OAU, AMCEN, and NATO, to mention a few, seek advice from experts to enable them to develop strategies for international environmental rule making or negotiations for international environmental laws and policies that are to be considered. Other nongovernmental organizations seek experts to play a role in the decision-making process of national policy development or in rule making as national governments develop environmental regulations and standards.

NGOs have become some of the most active driving forces in environmental management and the protection of natural resources. NGOs have indeed become more influential in shaping environmental management and protection in every country, and their increased activities are among the forces that provide markets for environmental consulting in every nation. The creation and growth of international environmental markets are in most cases due to NGOs' activities, which demand and force legislators and governments to prioritize environmental concerns and address environmental problems.

NGOs may or may not have the necessary technical environmental capacity and human resources, thus they tend to depend on environmental experts to perform critical environmental and scientific analytical activities. Opportunities created by NGOs for international environmental consulting are among the major driving forces in the international environmental market. The surge of NGOs in recent times has created forces that shape the demand for environmental consulting in both developed and developing countries.

NGOs have become environmental advocates, putting extensive pressure on national governments and international organizations to include environmental protection as a high priority on the national and global agendas. In the West, NGOs continue to seek expert environmental consultants to assist them in assessing the impact of pollution and management of waste products from industrial generators, in an attempt to seek protective regulations and standards for industries. Nonprofit organizations have also become partners to the multilateral and developing assistance organizations in implementing environmental projects, mainly in developing countries. The activities of the private business sector in ensuring sound environmental practice include their efforts in scientific investigation, research, and the assessment of environmental management of their industrial and business activities to minimize waste generation and to find avenues for escaping burdensome regulations. These activities, such as environmental remediation, are forces that also create opportunities for environmental consulting.

Examples of nongovernmental organizations include such groups as the International Council of Scientific Unions (which deals with international biological programs) and the International Union for Conservation (which deals with ecology, species survival, natural resources, parks, education, and environmental policy). Appendix VI provides additional information on nongovernmental organizations and international nonprofit organizations of possible interest to the prospective environmental consultant who is planning to enter the international marketplace.

2.2.4 International Donor Community

The international donor community plays a key role in shaping the environmental market in developing countries. For example, the Global Environmental Facility (GEF) was created and is administered by the World Bank, the UN Development Program (UNDP), and UNEP to assist less developed countries with a per capita income of less than $4,000 to tackle environmental problems within their countries. It supports research and gives technical assistance in various environmental activities such as those addressing pollution prevention, ozone depletion, biodiversity, climate change, and water pollution.

These international financing institutions help to shape the markets for international consulting in the developing countries. The international donor community's large investments in developing a nation's institutional capacity and strengthening its economy are forces that open opportunities for international environmental consulting.

The global market for environmental consulting is growing very fast in middle- to low-income countries in Africa, Latin America, Asia, and the West Indies. These developing countries are highly dependent on the international donor community to finance their environmental management programs.

2.3 GLOBAL ENVIRONMENTAL INDUSTRY AND MARKET

The environmental industry comprises the market segment and the environmental consulting sector. The market segment includes:

- Water resources management
- Water/wastewater treatment
- Solid waste management
- Hazardous waste management
- Air pollution control
- Environmental remediation (air, water, soils, and groundwater)
- Laboratory testing and analysis

The environmental consulting sector includes:

- Environmental research and development
- Development of policy and regulations
- Implementation and/or application of these components

There are many ways in which the environmental industry is viewed. Some may view it as a combination of environmental technologies with environmental management. Others may view it as an industry comprising various activities such as development of environmental policies and regulations, enforcement, pollution control, pollution prevention, remediation and restoration, and assessment and monitoring. This book defines the environmental field to be inclusive of all these factors. For the purposes of looking at the global environmental field, the focus here will be on the environmental technology industry and the various environmental sectors that create goods and services.

Increasing consumer awareness worldwide has played a major part in the development of environmental management, human and on-site workers' safety, and product safety. This has resulted in increased interest in environmental issues by multinational companies. The growth in activities and trading blocs within the international marketplace has increased international environmental markets and opportunities for consulting, as illustrated by the international trade activities in and between Asia, Europe, and the Americas. Figures 2.3, 2.4, and 2.5 and Table 2.3 present an outlook for the potential for international environmental markets. In particular, the 1992 Rio Earth Convention set the scene for major opportunities in international environmental sectors that were recommended under Agenda 21. These include sectors such as sustainable development, environmental technology cooperation and transfer, and global ozone/air quality consulting.

The global environmental market has indeed surged in recent years (since 1992) owing to global environmental concerns and the socioeconomic and political implications of natural resource management and preservation. The markets for environmental policy and capacity building grew rapidly in the early 1980s but have since stagnated because most of the international laws/treaties and in-country environmental legislation have made great-advances. Markets for consulting on municipal solid waste and hazardous waste management for developing countries have shown increased prospects, and opportunities are emerging in virtually all developing countries. In developed

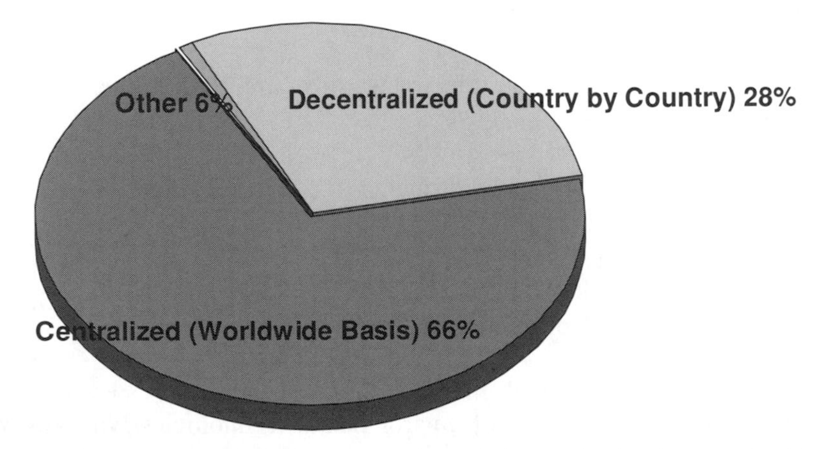

Figure 2.3 International environmental management by multinational companies (*Source:* December 1992 report by the McIlvaine Company)

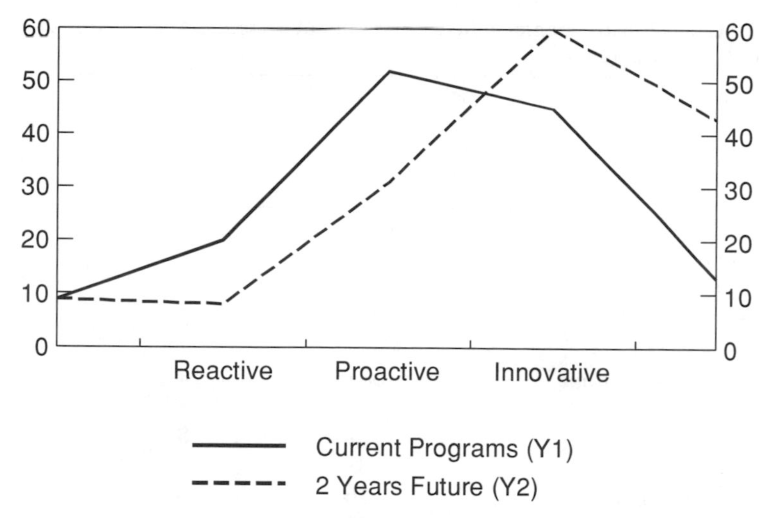

Figure 2.4 Multinational companies driving environmental markets through innovation (*Source:* Modified from Booz-Allen & Hamilton, *Columbia Journal of World Business* [Fall/Winter 1992])

countries, however, markets for municipal solid waste and hazardous waste consulting and environmental design and construction remained slightly stagnated in the mid-1990s, especially in 1994. It appears that many areas of environmental markets such, as environmental consulting (air quality, water quality, waste/toxic sites remediation, and wastewater consulting and engineering), were impacted and have consequently encountered some slow-down in Western Europe and the Americas, especially in countries such as the United States. Eastern Europe is an emerging market with tremendous opportunities for the out years in environmental remediation and preliminary

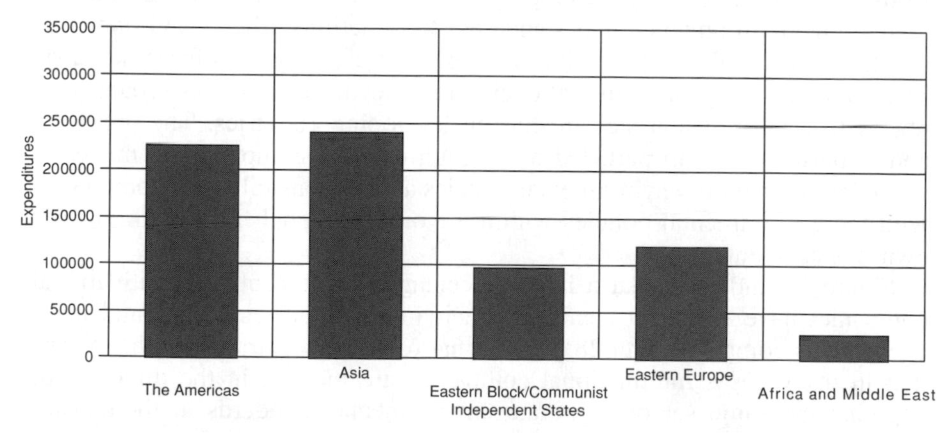

Figure 2.5 World power generation and air pollution control orders 1993–2003 (*Source:* December 1992 report by the McIlvaine Company)

Table 2.3 Global Environmental Industry and Market: Average Annual Growth Calculated over Five Years

Country	1992	Growth (%)	1997 Projected Growth (%)
United States ($180)	$134	6	10
Canada ($17)	$10	11	12
Mexico ($2)	$1	15	16
Latin America ($10)	$6	12	13
Western Europe ($132)	$94	7	8
USSR and Eastern Europe ($27)	$14	14	16
Japan ($31)	$21	8	12
Australia ($5)	$3	9	10
Southeast Asia ($13)	$6	16	18
Africa, Caribbean, and the rest of the world ($9)	$6	8	15

Modifed from *EBI*, Vol. VIII No. 8, Aug 1995. Environmental Business International Inc. Publishers, San Diego, California.

assessment, including environmental impact assessment consulting. It appears, therefore, that the greatest opportunities for Western hemisphere environmental experts and companies are in developing countries such as those that lie within the sub-Saharan African region.

For the past 10 years international multinational companies have come under criticism for not being environmentally responsible in business activities in developing countries. The situation in the developing countries stems from the notion that many developing countries lack adequate environmental regulations, requirements, and standards. In situations where environmental regulations exist, there are inadequate tools and capacity to enforce the laws. Therefore, in past years multinational companies have been less proactive in regard to environmental stewardship in developing countries. The pressures and criticisms have, in part, led many multinational companies to move toward development of environmental policies and responsibility for compliance with the environmental policies within the organizational framework of their own management.

Figure 2.3 indicates that a large percentage (66 percent) of multinational companies have shifted to centralize their international environmental management, as compared with 28 percent that remain decentralized. This means that in many cases multinational companies are moving in the direction of applying the same set of corporate environmental standards at their headquarters. In many cases, the headquarters is located in developed countries where there are stringent environmental requirements and therefore, corporations are subjected to higher standards of environmental laws and regulations. The shift toward environmental compliance by multinational companies, and, moreover, national companies realizing the public pressure and criticism and the increasing stringency of environmental regulations, has forged the movement toward innovative and cost-benefit approaches to environmental stewardship, better public relations, and promotion of corporate image.

Figure 2.4 illustrates the level of innovativeness of corporate executives and their respective companies. Such innovation is also driving environmental markets and the need for environmental consultants worldwide.

Figure 2.5 shows that the industrialized countries that have stringent environmental regulations and policies in place are experiencing higher expenditures on pollution abatement mandated by pollution control orders. This is also due to the emergence of the new order of trading blocs in these regions, which underscores the direction in which international business is taking shape relative to environmental pollution abatement. The middle-income and developing countries have not yet secured stringent environmental regulations and policies, therefore expenditures and market for pollution control equipment orders are stagnant or very low. The future for developing countries to achieving greater pollution abatement is promising as industrialized nations increase their technical assistance and financial aid to developing countries for pollution abatement and sustainable development.

Table 2.3 provides data on the growth of the global environmental industry. It is clear that the public demand for environmental remediation and a stringent environmental policy framework is driving environmental industry growth. Although developing countries show a moderate growth percentage, their demand for environmental technologies and management will increase at a faster rate than that of the rest of the world by the year 2000. The African continent is a typical example of a potential growth region for the environmental market, simply because of the ongoing dynamics of the region's environmental protection and management activities, such as:

- The development and implementation of the National Environmental Action Plans (NEAPs)
- Increasing development and creation of environmental ministries and departments within the governments
- Development of environmental statutes, regulations, and standards
- Multilateral corporations in Africa adapting environmental standards established in their countries of origin
- Increased international and local in-country "grass roots" pressures to protect and manage natural resources and the environment
- Increased public awareness in African communities

2.3.1 Global Environmental Market Activities

The global environmental market is driven by various activities, such as:

- Environmental policy and regulations
- Environmental technologies industry
- Export and trade programs by international multilateral financial institutions and governments

Environmental policy and regulations. The increasing actions and activities of grass roots organizations, nonprofit groups, international governing bodies, and governments have led to the institution and creation of governmental departments and agencies mandated to develop environmental regulations and implement the requirements of the regulations to protect the environment and natural resources. These activities have tremendously increased, and will continue to increase, the demand for environmental consulting, especially in developing countries where environmental expertise and technologies are lacking.

Environmental technologies industry. The increased research and development to meet ever increasing environmental regulations and standards have led to a growing industry for environmental technologies. The need for corporations to reduce the costs of pollution control disposal, treatment, and compliance has played a major role in the global environmental market. The approximate growth in the world market is estimated at $200 billion to $300 billion annually as reported by the OECD and EBJ, and it is estimated to sustain a steady growth beyond the year 2000.

The United States currently leads in environmental technologies and thus forces the increase in and demand for U.S.-based environmental experts in international environmental consulting. Table 2.4 charts the trading of various environmental products by leading industrialized nations (G7 nations) in the environmental technologies industry.

In one sector, predictions by the Mcllvaine Company indicated that the global air pollution control industry would generate 10 percent more orders in 1989 than in 1988, as shown in Figure 2.6. The worldwide air pollution control industry is expected to show increments of approximately 1 billion dollars per annum of air pollution control orders for the next 10 years. The driving forces for the market are several basic factors, such as the opening of the global economy and international markets, the growing gross national products of major industrialized countries such as the United States and Germany, the global accords on ambient air quality, and the requirement by the global community to reduce air pollution. The global accords (such as the Montreal Accord), in addition to in-country pressures in regard to air quality, have led to more stringent regulations and standards that have, in turn, led industries to retrofit their air pollution control equipment to meet current

Table 2.4 Trade in Environmental Management Products and Technologies

Country	Total Output (U.S. $ Billion)	Percentage Output (U.S. $ Billion)	Trade Balance (U.S. $ Billion)
	United States		
OECD (1990)	80	10	4
EBJ (1992)	134	5	N/A
	Germany		
OECD (1990)	27	40	10
EBJ (1992)	36	31	N/A
	Japan		
OECD (1990)	30	6	3
EBJ (1992)	21	23	N/A

Source: OECD Environmental Industry Situation, Prospects and Government Policies, Paris 1992. *Environmental Business Journal* (EBJ) 1995, Environmental Business International Inc. Publishing Co., San Diego, California.

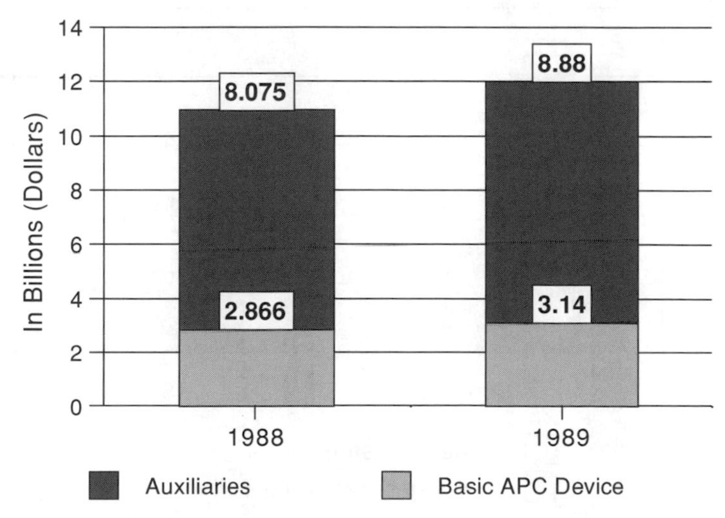

Figure 2.6 World air pollution control market (*Source:* R. W. McIlvaine, Air and Waste Management Association, Air and Waste Management Journal, 39, No. 3 [March 1989])

emission and performance standards. In the Western hemisphere, the industrial boilers and utility sector is most affected by the growing demand for air pollution control equipment. For example, in Germany the industrial boilers and utility sector outpaced all other industries in air pollution control orders between 1988 and 1989, as shown in Figure 2.7. This trend is likely to maintain its momentum and growth for the next 10 years.

Export and trade programs by international multilateral financial institutions and governments. The trend toward globalization of the world economy, especially in environmental export technologies, presents unique challenges for governments and international development and financial organizations, and the need for countries and international institutions to assess and understand the economic, political, and market factors such as:

- Country-specific legal, political, and enviro-economic trade regulations and structures
- Market access and trade barriers
- Availability of export financing for environmental technologies

The United States has instituted several programs within different governmental agencies and departments, such as the U.S. Department of Commerce, to assist U.S. companies. Programs administered by the U.S. Commerce Department and other departments of government include:

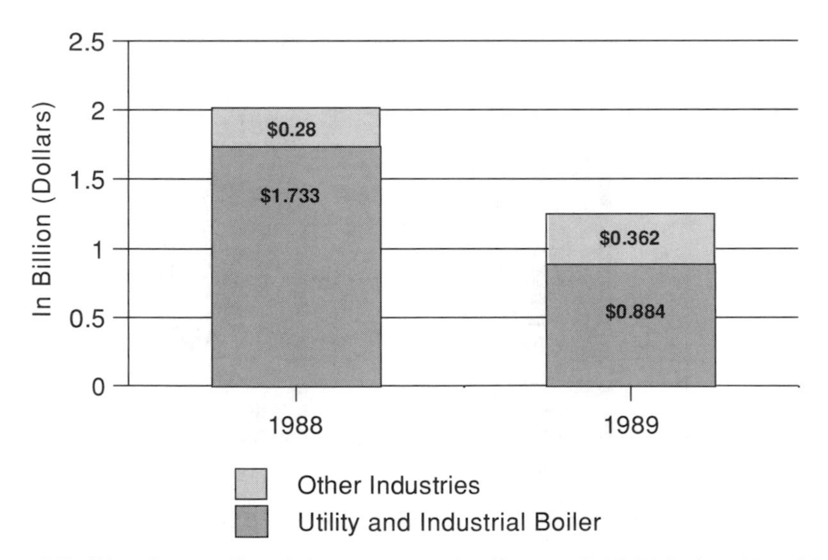

Figure 2.7 West German air pollution control market (*Source:* R. W. McIlvaine, Air and Waste Management Association, Air and Waste Management Association Journal, Pittsburgh, PA, 39, No. 3 [March 1989])

- International Trade Administration
- National Data Bank
- Exports Contact List Service
- Foreign Buyers Program
- Overseas Private Investment Corporation's Overseas Investment Opportunity Program
- The United States Agency of International Development (USAID) Center for Trade and Investment Service Program
- U.S. Department of Energy's Overseas Environmental Technologies Demonstration Program
- The Small Business Administration's Office of International Trade Assistance Program

The European Community (EC) has similar programs within its member countries and within its activities to encourage environmental technologies within its member countries, for example, the Integrated EC Financing Programs under EC's financial instrument for the environment (LIFE) Program. LIFE program is aimed at the conservation of natural habitats and of wild fauna and flora of European Union (EU) interest. EC provides financing through LIFE for technical support and other services outside EU community territory (LIFE-Third Countries).

The international financial and multilateral development institutions, such as the Asian Development Bank, the European Bank for Reconstruction and Development, the Inter-American Development Bank, and the International Bank for Reconstruction and Development, among others, have increased their activities and priorities in several programs and financing for environmental technology exports and environmental consultancy for environmental markets and services.

2.4 ENVIRONMENTAL TECHNICAL ASSISTANCE NEEDS IN DEVELOPING COUNTRIES

Developing countries face numerous environmental problems that threaten their long-term growth and socioeconomic stability. Urbanization, poverty, population explosion, and migration are contributing factors to the widespread environmental deterioration in developing countries. Poverty and the inability of local governments to manage the physical infrastructure and basic services have led to declining human health conditions, overcrowding, increased exposure to toxic and hazardous wastes, unsustainable agricultural practices, and the growth of squatter settlements in ecologically sensitive areas.

Industrial growth has been on the rise since the early 1960s in most developing countries, (especially in Africa), where environmental management has been minimal—a situation that has contributed significantly to environmental deterioration and associated human health problems. Developing countries are increasingly in need of environmental technical assistance and technology transfer to address the wide range of environmental problems such as:

- Reducing long-term threats to the environment, resulting particularly from the loss of biodiversity and the possibility of rapid climate change
- Designing and implementing sustainable development
- Designing and implementing a comprehensive municipal solid waste management program
- Providing environmental training and assistance to governmental institutions and capacity building

Population growth and infrastructure development in developing countries constitute an area in which technical assistance and international environmental consultants are in demand. For example, in sub-Saharan Africa, the population is growing at an alarming rate: from 500 million in 1990 (World Bank, 1992), to an estimated 1.5 billion by the year 2030. The expanding population, especially in urban areas, is placing extreme stress on ecological systems in developing countries. The population growth, compounded by inadequate socioeconomic and environmental policies and lack of expertise in the region to provide basic infrastructure and social services, has led to massive and increasing environmental degradation. The problem is illustrated in Table 2.5, which shows the access to different types of primary sanitation systems (which are inadequate or need redesigning or construction) for some cities in sub-Saharan Africa.

Global response to the environmental challenges faced by developing countries has resulted in the availability of global funds and some financial resources to governments in the region to provide for basic infrastructure services such as sanitation, appropriate sewage treatment, and conducting environmental impact assessment and feasibility studies for economic development projects. Developing countries lack the technologies and expertise;

Table 2.5 Primary Sanitation Systems in Selected Cities in Sub-Saharan Africa

	ACCRA	ABIDJAN	DAR ES SAALAM	OUAGADOUGOU
Pit latrine	41		55	88
Pan toilet	20			
Septic tanks		55	25	
Flush toilet	35	30		
No access (Open defecation)	4	15	20	

Source: GREA, "*Plan Strategique de'lassainiissement des eaux uses des la ville de Ouagadougou* 1993." *Urbanization and Environment in Sub-Saharan Africa,* World Bank, 1995.

therefore, they turn to international experts in addressing the environmental problems that confront them.

The need for economic development assistance is on the rise in developing countries. For example, the sub-Saharan Africa (SSA) region alone accounts for 21 out of the 30 poorest countries in the world, with negative per capita income growth rates. The increase in developmental activities has revealed new technical issues—basically, the synergies between development and environmental protection that are encountered by industrialized countries. These rising issues have led to increased need for technical assistance by developing countries and by the international donor communities. Since the latter part of the 1980s, the region has experienced a decline in per capita income. Most SSA countries have embraced the Structural Adjustment Program (SAP) of the World Bank and programs of the International Monetary Fund (IMF) and have undertaken considerable economic reform activities. These economic development programs reflect the new issues that have added to urban waste management and environmental problems.

The complex dynamics of environmental issues and the lack of technical environmental experts and environmental technology in developing countries have led these countries to depend, to a large extent, on consulting assistance from the industrialized countries. The environmental business opportunity market in developing countries is expected to grow rapidly. Municipal solid waste management provides a promising future, especially in the area of landfill design and construction. Water pollution control and hazardous waste management will gradually increase in market size by the year 2000. Figure 2.8 provides a forecast for the year 2000 for major areas for environmental market opportunities in developing countries.

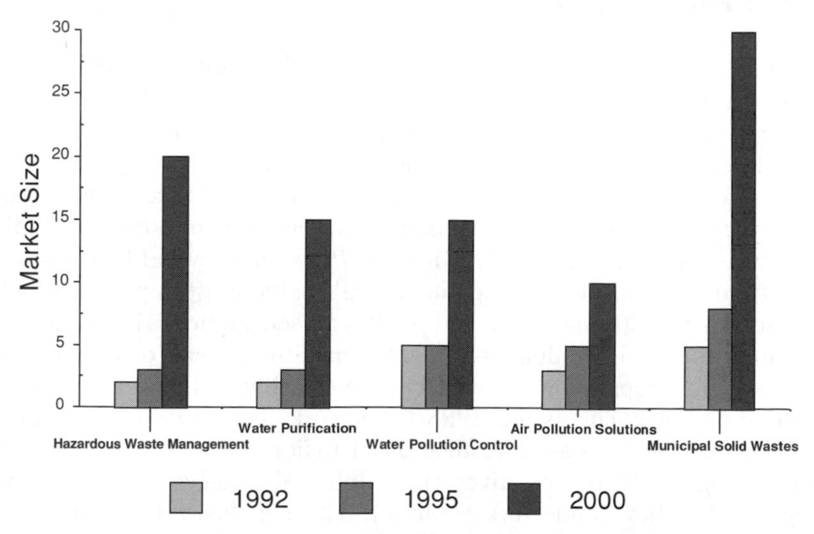

Figure 2.8 Environmental business opportunity markets in developing countries

3

PREPARING FOR AN INTERNATIONAL ENVIRONMENTAL CONSULTING PRACTICE

It is important for one to develop a responsible strategy in preparing to enter the international environmental consulting practice. The framework shown in Figure 3.1 can be used as a guide. This chapter provides in-depth information, including tools and strategies to help prepare individual consultants, nonprofit environmental organizations, and private consulting firms to enter the field.

3.1 THE FUNDAMENTALS

Starting an international environmental business is unique and challenging in comparison to starting a domestic business. The principles of starting a domestic business in many ways do not necessarily transfer to the international marketplace, inasmuch as the latter requires consideration of several intricate factors. A carefully designed plan of action to begin selected activities is necessary to ensure a smooth, methodological entrance into the marketplace. This plan is called the *Entry Preparation Plan Strategy* (EPPS). There are several factors to consider in preparing an EPPS, the most important of which are those that fit with the vision and goals you set. Figure 3.1 presents some major areas for the individual consultant, consulting group, or corporation to consider in the preparation stage of entering an international environmental consulting practice. This is the EPPS framework. The EPPS is a master plan that comprises the strategic investigation of major and minor factors that aid in identifying strategic objectives (i.e., financial strategy, human resource strategy, and policy framework strategy). The conclusions and findings of

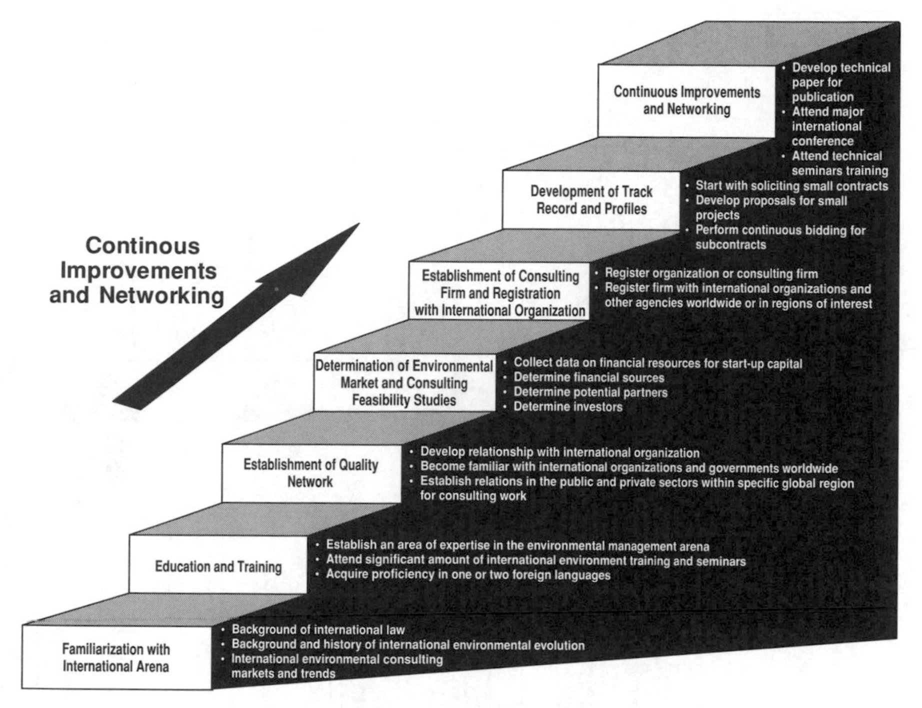

Figure 3.1 Framework for international environmental consulting (*Source:* P.A. Sam, AERCG 1996)

strategic investigations provide a directive for entering the international consulting market and for the longevity and survival of the organization. Figure 3.2 provides an example of a broad EPPS for each category of entry. Apart from the suggested steps provided in each category, there are other factors that must be considered for a broad EPPS:

- Evaluate yourself and your company/organization.
- Take a preliminary inventory of your capabilities and tools.
- Visualize success and develop an action plan for success.
- Anticipate pitfalls and develop strategies to overcome them.

Evaluate yourself and your company. Without a holistic evaluation of yourself or your company prior to entering an international environmental consulting practice, it will be difficult to effectively control activities to meet your goals, objectives, and mission. The international marketplace, as mentioned previously, is dynamic and presents various socioeconomic, political, and technical challenges. A consulting enterprise must carefully evaluate its own functions, attributes, and characteristics in order to meet the challenges that present themselves during the life of the business. The paradigm for

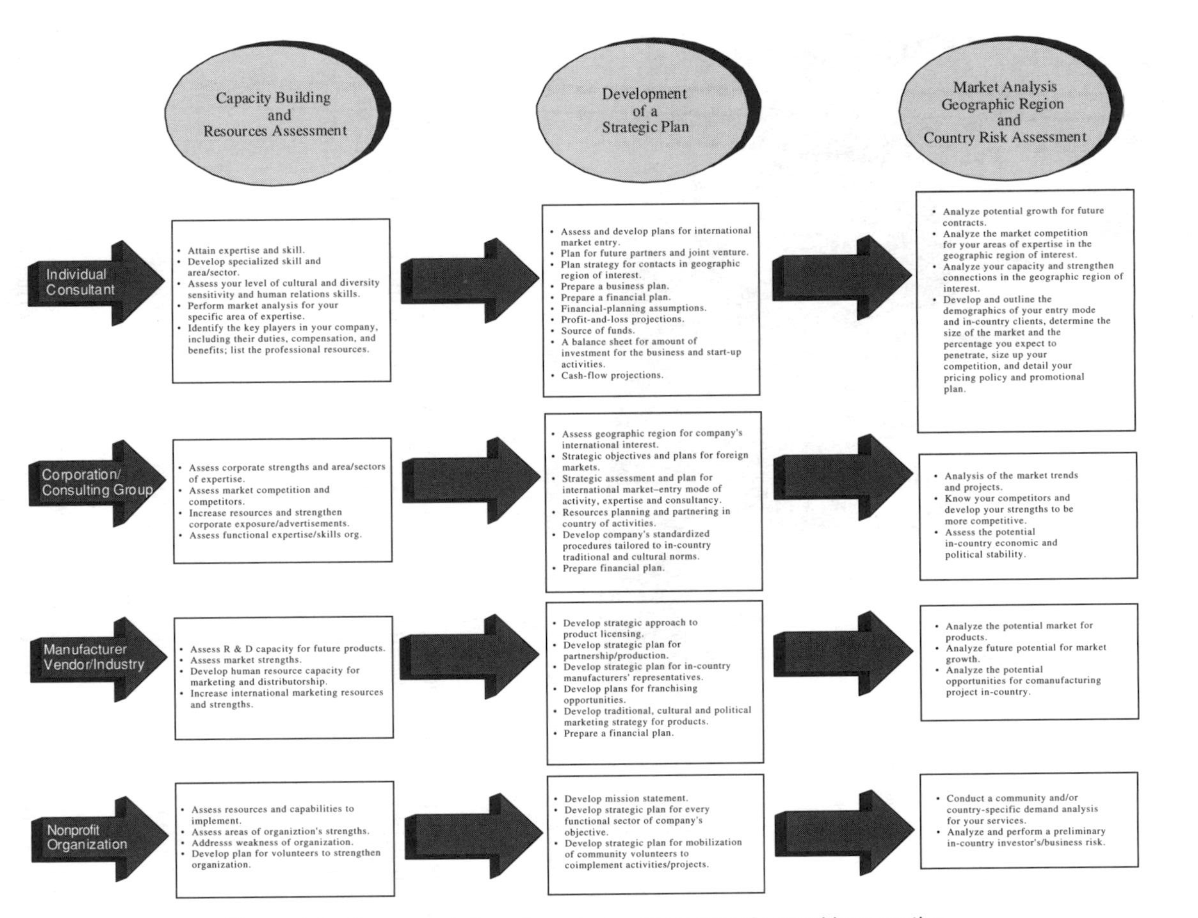

Figure 3.2 Suggested steps for entering international consulting practice

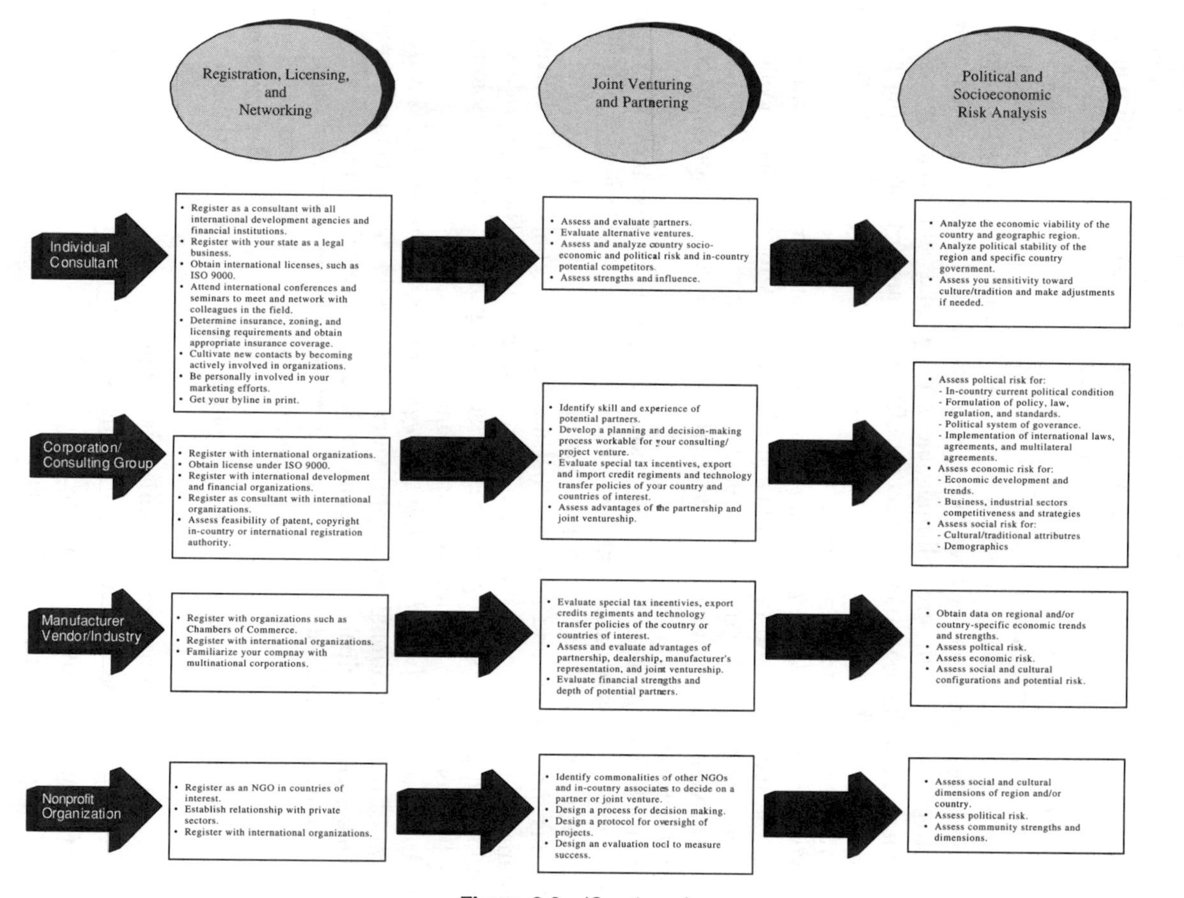

Figure 3.2 *(Continued)*

knowing yourself is a series of self-evaluations and continuous accounting. It is built on understanding yourself and believing in your mission. This is a first and essential step to success in the international arena. Jacob M. Braude wrote: "Any person who wants to succeed must begin by believing wholeheartedly in his own ability. [He] cannot expect others to believe in him unless that person believes in himself, for it is a cardinal rule of life that the world takes a person at his own evaluation."

Take a Preliminary Inventory of Your Capabilities and Tools. Knowing the strengths and weaknesses of yourself or your company provides opportunity for positive readjustments and quality management. It offers a gateway to effective organization, streamlining, and setting goals for yourself or your company in utilizing resources in a cost-effective manner to meet your goals, objectives, and the mission that you have set forth in entering the international environmental consulting practice.

Visualize success and develop an action plan. You can proceed with planning for success in the international marketplace. After you have evaluated you current potential, your capabilities and tools, and your situation relative to what your are planning to do, you are now ready to develop your action plan for success.

Before developing an action plan, however, you must first have a mission statement. A mission statement is the cornerstone for developing and designing an action plan. An action plan simply consists of step-by-step tailored actions, procedures, and activities that are based on goals and objectives to attain a successful international environmental practice. These provide well-engineered processes and guidelines for achieving the optimum mission that has been outlined in the mission statement.

Anticipate pitfalls and develop strategies to overcome them. The dynamism of the international market provides surprises in every sector. In many instances an unanticipated event can interfere with your time line to complete an activity or totally disrupt the objectives of your action plan. Therefore, you have to develop a contingency plan to anticipate pitfalls and tackle unforeseen circumstances. Most obstacles emanate from external factors, for which you should prepare contingency plans. These are a few examples:

- Constraints on financial resources and capital
- Changes in the market (need to adjust your initial market entry mode)
- Changes in political conditions
- Changes in economy and infrastructure

As the world moves into a new era of environmental management, protection of the earth's natural resources, and globalization of environmental management, the environmental profession is beginning to respond to this challenge by shifting into a globalized field of practice. An increasing number of individuals, consultants, and corporations are preparing for the vast opportunities that exist internationally. It is important for consulting entities to plan, study, and design preentry strategic plans and acquire the necessary information and added skills to furnish them a competitive edge and/or advantage in this rather complex and competitive market.

The most important aspect of international environmental consulting practice is the need for practitioners to be sensitive to different cultures and diversities. Like that of most businesses, the cornerstone for a successful international environmental practice is to maintain principles reflecting ethical values and morality. This set of principles requires you to know and understand your obligations in your practice. In addition, several questions come to mind in beginning a new profession or field of practice such as international environmental consulting.

3.1.1 Understanding the Field

Before you enter the global marketplace for environmental practice, you must know what it is all about by asking Who? What? Why? and For whom? The answers to these questions should provide a better understanding of the field of consulting and help to build the ethics of practicing international environmental consulting.

1. *Who* is an international environmental consultant?
2. *What* does an international environmental consultant do?
3. *Why* consulting?
4. *For whom* do you consult?

Answers

1. *An international environmental consultant* is a person or entity, such as a corporation, that provides environmental consulting to various clients (in both the private and public sectors) in a foreign country.
2. *What does an international environmental consulting practice entail?* It comprises a general provision of technical and nontechnical assistance through thoughtful interventions of different sorts, such as:
   - Application of theories and principles
   - Confrontational approaches
   - Prescriptive approach
   - Catalytic approach
   - Acceptant approach
3. *Why consulting?* The consultant does the job of consulting to resolve issues and design remedial approaches with the following intentions:
   - To define goals and objectives of a problem
   - To normalize and standardize situations

- To provide an avenue for power and authority towards resolution, and to meet the needs and wants of the client

4. *For whom?* The consultant serves the needs and wants of the clients who may fall into any of the following categories:
   - A government entity
   - A business entity
   - A group (e.g., a community)
   - An organization (e.g., profit or nonprofit entity)
   - An individual entity

3.1.2 Understanding the Practice of a Consultant

Understanding the preceding descriptions and definitions is a *first step* in developing an understanding of international environmental consulting and in the process of developing a *vision* and a *focus* on the dynamics of international environmental practice. The general process follows the principles of FADE, which stands for Focus, Analyze, Develop, and Execute. The second step in the process is *goals setting.* History shows that the most fundamental principle in any human endeavor is to have a goal. In preparing to enter the international environmental consulting practice, individuals and companies must set goals. What must you consider in establishing and setting goals? Before you proceed with any plans, you must have a vision. From the vision comes the mission, and from the mission come specific goals and even objectives. The rationale for establishing a goal is usually based on individual style, abilities, ideology, and personality. The goals-setting process assists individuals and firms planning to enter the international environmental marketplace to identify:

- Short-term (one- to two-year) need for entering the international marketplace, met by improving current capabilities and performance
- Long-term (three- to five-year) business strategies, opportunities, and options

In the present era of corporate restructuring, many corporations, through the principles of Total Quality Management have established their goals based on their corporate characteristics, structures, and mission. Goals must be tailored to the end product and/or vision for the future. Goals can be visualized as an emergency survival kit for present and future operations and to measure future success and reward. Some of the elementary basics for establishing goals are listed in Table 3.1. Each of these basic elements should have the following characteristics:

- Personal to the individual or organization

Table 3.1 Some Basic Elements to Consider in Establishing Goals

• Expertise and talent	Types of client interaction
• Contribution to profit	Productivity of employees
• Enhanced communication	Employee participation
• Market share	Customer satisfaction
• Quality and service	Innovation and creativity
• Strategy formulation	Problem solving
• Employee assistance	Research and development
• Motivation and morale	Priorities/time management
• International market forces	Cultural diversity
• Design-to-market timing	Safety, international agreements/law
• In-country regulations	Financing

Adapted from Alan Weiss, *Million Dollar Consulting* (McGraw-Hill, New York, NY), 1992.

- Challenging and achievable
- Specific and measurable
- Compatible with mission statement, values, priorities, and existing obligations

3.2 INTERNATIONAL CONSULTING PRACTITIONER'S QUALIFICATIONS, CAPACITIES, AND ABILITIES

Upon successful establishment of goals, the next step in the process is to examine the individual's or company's qualifications, capacities and abilities for effective implementation of business strategies and objectives. In addition to adhering to the basic framework for preparing to enter into international environmental practice that was presented earlier, a careful, methodological evaluation of capacity in the following areas should be undertaken and an honest decision made as to whether you or your firm has what it takes to enter this practice. Conducting such an evaluation will unveil any areas of weakness that may exist. The areas of weakness that are evident must be addressed, and steps taken to build and strengthen those areas. The evaluation must include critical questions that should be asked and analyzed for the individual or the company, such as:

- Do my educational background and credentials identify me as a specialist?
- Do I have adequate education and an advanced college degree in the field?
- Would my personality and personal life, including family life, infringe on my practice as an international environmental consultant?
- Am I tolerant of different cultures and traditions?

- Do I have general orientation to the workings of international business, trade, and other activities?
- Do I have experience in sole proprietorship?
- Does my company have a successful record in delivering services and customer satisfaction?
- Does my company have the human resources, capacities, and abilities to undertake the unique and dynamic international environmental tasks?
- Does my company have the financial resources to expand into the international marketplace?
- Does my company have any networks in the international arena?
- Have I ever traveled overseas, and if so, am I comfortable with long hours of traveling in addition to long working hours?
- Do I speak more than one language?
- Have I developed and published technical papers?
- Do I have sufficient living expenses and savings for personal and business expenditures for one year?
- Do I have the technical skills, expertise, and specialty with a good track record extending over a few to several years?
- Can I handle and deal with extreme stress and pressures?
- Do I have a set protocol for my international environmental consulting practice?

Note: If you honestly and strongly believe that your answers to the preceding questionnaire are satisfactory, you are ready to prepare and plan to enter the international environmental consulting practice.

Now that the evaluation of yourself or your company has been completed, you have become aware of areas of weakness that must be addressed.

3.2.1 Addressing Your Weaknesses

The most important aspect of preparing to enter the international environmental marketplace is to strengthen your capabilities, abilities, and your organization structure for optimal performance and efficiency. Knowing your weaknesses is your strength. Therefore, upon critical evaluation of your qualifications, capabilities, and abilities, you should be able to identify your weaknesses and address them by increasing the capacities of areas of deficiency and building or strengthening organization structures that are weak.

A basic approach for addressing weaknesses is to examine the characteristics of the identified areas of concern and evaluate the management effectiveness for each. Characteristics and dimensions of management effectiveness include, but are not limited to, the following:

- Technical competency
- Communication effectiveness
- Interpersonal sensitivity orientation
- Performance and action orientation
- Leadership
- Flexibility
- Vision and focus
- Sensitivity to the surrounding environment
- Strategic view
- Broad perspective of the individual and the organization

Technical Competency: Specialized human resources or expertise such as engineering, environmental science, accounting, and competencies in management and operations, as listed in Table 3.2.

Communication Effectiveness: The ability to communicate effectively with customers through verbal communication, writing, and listening.

Interpersonal Sensitivity Orientation: This area covers the knowledge base and awareness abilities or capabilities of the individual or organization to detect and address potential problems to the customer and the public at large. Other components include the strengths of the individual or the organization in effective negotiation, conflict resolution, persuasion, and marketing.

Performance and Action Orientation: This area includes critical attributes such as independence, proactiveness, risk taking, accepting responsibilities, and capability of being decisive.

Leadership: The ability and capability of the individual or organization to "take charge," to lead and manage effectively.

Flexibility: This area covers the strength and ability to be open to acquiring new information, behavioral flexibility of the individual or the organization, and a degree of tolerance and patience for ambiguities, unexpected changes, and stress.

Vision and Focus: The ability and capability to seek goals, meet objectives, and adhere to the mission statement; the ability to follow through and deliver on contracts/projects.

Sensitivity to the Surrounding Environment: The ability to be aware of the socioeconomic, political, and technical elements of the business environment in which the individual or company operates.

Strategic View: This area includes the ability to collect and analyze information and diagnose, problem solving capabilities, and the ability to reach a decision or render judgment.

Broad Perspective of the Individual and the Organization: This area consists of the clear definition of the mission, long-term goals and objectives, and short- and long-term considerations to sustain the business practice.

Table 3.2 Management and Operations Competency Areas and Functions

Competency Areas	Environmental Consulting Management Functions
I. *Integrating Internal and External Issues.* This competency area requires the individual's or company's effectiveness in staying abreast of external issues germane to internal activities and operations and applying this information in decision making and operations within the company.	1. *External Awareness.* Identifying and keeping up-to-date with key business issues, regulations requirements, and technological trends (e.g., economic, political, social, technological) that are likely to affect the consulting business.
	2. *Interpretation.* Keeping partners, key associates, and clients informed about critical policies, priorities, issues, and trends and how these are to be incorporated in project activities and products.
II. *Representing and Coordinating.* This competency area focuses on the effectiveness of external communications.	3. *Representation.* Presenting, explaining, selling, negotiating, and defending.
	4. *Coordination.* Performing liaison functions and integrating internal activities with the activities of other entities, including subcontractors, clients, and other business entities.
III. *Planning and Guiding.* This competency area includes effective management of activities and projects. It entails well designed and established goals, objectives, and priorities, and the structures and processes necessary to carry them out.	5. *Project/Activities Planning.* Developing and deciding on longer-term goals, objectives, and priorities, and developing and deciding among alternative courses of action.
	6. *Internal Guidance.* Converting plans to actions by setting short-term objectives and priorities, scheduling/sequencing activities, and establishing effectiveness and efficiency standards/guidelines.
IV. *Administering Financial Resources and Material Resources.* This competency area deals with responsible management of running capital and expenditures, allocating financial resources and material resources necessary to support projects.	7. *Budgeting.* Preparing, justifying, and/or administering the available budget.
	8. *Material Resources Administration.* Effective utilization of supplies, equipment, facilities; appropriate procurement/contracting activities; efficiency of logistical operations.

Table 3.2 (*Continued*)

Competency Areas	Environmental Consulting Management Functions
V. *Utilizing Human Resources.* This competency area covers responsibilities and processes for ensuring that personnel and talents are appropriately employed, effectively and efficiently utilized, and dealt with in a fair and equitable manner.	9. *Personnel Management.* Projecting the number and types of staff needed by the company or for a specific project, and using various personnel management system components (e.g., recruitment, selection, promotion, performance appraisal) in managing projects.
	10. *Supervision.* Providing day-to-day guidance and oversight of subordinates (e.g., work assignments, consultation, etc.) and actively working to promote and recognize performance.
VI. *Reviewing, Implementation, and Results.* This competency area consists of the ability to ensure that progress on projects is being made and deadlines are being met and that other elements such as policy, standards, and requirements of projects are being implemented. Finally, it includes the ability to adjust as necessary, to meet the goals and objectives of the projects.	11. *Monitoring and Quality Assurance.* Continuous oversight on the overall status of activities, identifying problem areas, and taking corrective actions (e.g., rescheduling, reallocating resources, etc.).
	12. *Self-Evaluation.* Critically assessing the degree to which activities/project goals are achieved and the overall effectiveness/efficiency of operations, to identify means for improving efficiency and performance.

Source: Peter A. Sam, 1996.

3.3 THE ACTUAL BUSINESS: HOW TO START

The next phase in the general preparation is the start of the actual business of international environmental practice.

Every international environmental practice starts from ground zero. The building blocks of every consulting practice, whatever the manner of entry, *must* be structured and built from a foundation of intelligent strategic planning. There are no set principles or theories applicable to preparing and setting up an international environmental consulting practice, but there are basic requirements and building blocks to lay the foundation before entering the marketplace. In any case, there are many reasons for an individual or a company to venture into new territories, such as expansion of business, tapping into new and promising future markets, and increasing profits.

Preparation and readiness to enter an international market reside solely with each individual and/or company. There are four steps in the process:

1. The planning process
2. Strategic thinking: Developing alternatives
3. Selection of optimal alternatives
4. Implementation of action plans and alternatives

3.3.1 The Planning Process

Planning is the first step in the preparation process, and you may need to seek outside professional advice in your planning activities. Planning is a way to enhance your prosperity and the success of your international environmental practice.

The planning phase includes but is not limited to:

- Data collection, analysis, and evaluation
- Risk management consideration
- Development of business plans
- Feasibility studies
- Strategic thinking

The most important aspects of the planning phase are the feasibility study and consideration of risk management factors, on which you should focus attention at all times throughout the planning process. Additional areas that you should address are discussed in the following paragraphs.

Preparation and Development of an International Environmental Business Plan. Developing a business plan brings into focus your business expectations and your entry into the market. A plan is essential for involving partners, potential investors, and financial institutions, should you need to borrow money or seek financial support. Although it can be expanded to include other applicable elements, the basic ingredients of the business plan must include the following:

- A description of the business
- A marketing plan
- A management plan
- A financial plan

The marketing plan becomes a tool to effectively market your services and reach potential international clients.

Registration of Your Business or Organization with the State or Federal Government Department. What are the business registration requirements by your state, city, or federal government for an individual or a corporation doing business overseas? Nongovernmental organizations (NGOs) must register with the federal government and inquire about registration in countries where they intend to practice and provide services.

Seeking Legal Advice on Foreign Transactions with Foreign Countries. In some cases there may be a trade embargo (or international sanctions) on a country in which you seek to undertake your environmental consulting practice. State and federal departments provide information and assistance in foreign business and international activities. Therefore, you should not solely hire and depend on an attorney. There are several self-help pamphlets developed by the U.S. Small Business Administration, the U.S. Department of Commerce, and the State Department.

Seeking Financial Resources for the New Market You Are Pursuing. Are there financial institutions that will provide loans and capital? Can your company invest in this venture with minimal risk? Are your financial resources adequate to survive at least two to three years after entering the international market—even when profits are locked in? Financial planning is especially important in pursuing international business ventures. In addition to the usual expenses incurred in domestic ventures, there are costs for overseas travel, international communications, and other items that you will encounter while in a foreign country. For this reason, you will not want to be stripped of business capital upon establishing your international consulting activities. Many businesses and individual consultants in the international market fail most often because of depletion of resources and financial constraints. Be financially wise: find ways, means, and techniques to minimize the cost of setting up your practice or commencing your international environmental business. Seek opportunities to enhance your accounting competence by applying appropriate accounting principles and obtaining pertinent information. These approaches will allow you to lower your financial risk through planning effectively and lowering overhead cost.

Assessing Investment Considerations. What are the calculated investment requirements, needs, and risks associated with your international environmental practice? Do you incorporate your company, seek shareholders' acknowledgment, sell new stocks to finance the new venture and/or seek partnership or co-investors, or seek government assistance? Preliminary assessment and feasibility consideration is a must. You will also need to answer these questions and seek appropriate assistance if the situation you confront warrants professional advice.

Determining Human Resource Capacity. Determine human resources capacity, expertise, and needs. Make adjustments to current structures in your company to optimize your expertise and provide the necessary staffing capacity to perform and optimize your technical, marketing, and administrative competence.

Assessing Insurance Coverage. Employee compensation insurance, professional liability insurance or quality assurance guaranty, company insurance, comprehensive insurance, workers' compensation, life insurance, health insurance, and general liability coverage for you and/or your corporation are very important. The type of insurance you or your company needs depends on the type of practice and sector of the international environmental field. If you already have insurance and are preparing to increase your coverage for your international expansion, in many cases your insurance costs will change as a result of activities; cost depends on the coverage you want and to what extent you will need that coverage. It is very important to have as much coverage as possible. The high cost of mistakes has led consultants to acquire reasonable insurance coverages. Liability protection is a must, because incidents such as personal injury, negligence, and/or property damage can result in costly claims that can be levied against companies, manufacturers, and individual consultants in the wake of malpractices. Some foreign environmental projects and/or contracts require certain amounts of insurance coverage. There are other types of coverage, such as business interruption insurance, which is vital for an international consulting business. Owing to the uncertainties of governments in the developing countries, you may face a sudden halt in activity because of a socioeconomic or political situation. You may have to get advice on what insurance coverage you or your corporation will need for the specific activities and consulting you are preparing for and practicing. Insurance coverage in many instances reduces the cost of unexpected externalities, which eventually eradicates the risk of transaction cost differential that in many instances drives individual consultants and small- to medium-size consulting companies out of business. Insurance coverage for investment is essential for small businesses and individual consultants. This coverage insures investments in countries with unstable political and economic conditions against risk of currency inconvertibility, political violence occurrences, revolution, insurrection, and expropriation (owing to loss of an investment as a result of nationalization or confiscation by the in-country government). Without coverage, incidences could be fatal to your consulting business.

Considering, Assessing, and Procuring Services and Contracts. One of the most important questions to consider in planning activities is how are you going to attract clients for consultancy and contracts? How are you going to plan for marketing your services? How are you doing to sell your response to proposals (RFPs)? There are many ways to seek contracts and services for international environmental consulting and business. Chapter 5 of this book discusses contract bidding, proposals, and negotiating, telling individuals and companies preparing to enter international environmental practice how to develop appropriate proposals and bids and gives information on the pricing of services and procuring services.

3.3.2 Strategic Thinking: Developing Alternatives

An international environmental market strategy must be developed to assess the current situation and the likelihood of future opportunities. A comprehensive marketing plan allows you to develop alternative and contingency plans to tackle the potential risks that are present in the foreign marketplace. There are a number of factors that fall within the domain of external environmental factors mentioned earlier:

- International laws
- In-country trade and marketplace attributes, environmental policies, and governance
- Various bureaucracy and governmental functions and structures
- Cultural and traditional configurations and practices
- Socioeconomic and business configurations and forms of transactions
- In-country marketplace attributes, activities, and sensitivities
- Complex forms of registration and licensing

International Laws. There are several international agreements and legal instruments relating to environmental concerns. Most developing countries are unhappy about the terms of market access for goods and services and areas such as intellectual property, investment, and consulting services. Opposition by most of these countries leads to the adoption of different trade terms. Table 2.1 of this book presents current international environmental agreements and laws. It is important also for a consultant to check with the in-country embassy to obtain information on trade, commerce, and environmental laws. It is also important for the consultant to check with his or her own government's Commerce Department, State Department, or Department of Foreign Affairs to obtain information about trade, tariffs, and other laws that may be applicable to doing business in a specific country. Other trade agreements and legal instruments may also exist. These trade, tariff, and environmental laws must be complied with. The minimum standard established under a specific treaty provides the basic set of rules that a signatory or a nation must establish. This provides an equal footing for all parties, but parties to a particular treaty can develop more stringent standards. Therefore, the consultant must be aware of these laws as well. Moreover, implementation of these international agreements and laws is sometimes interpreted differently by different countries, and any infringement may lead to fines and disbarment by government agencies in countries where the incident occurred.

In-Country Trade and Marketplace Attributes, Environmental Policies, and Governance. Each country has its own dynamics, attributes, and value system. Among the most important attributes of a country is its type of government, inasmuch as the government establishes trade and environmental policies. Almost all countries have their independence and are members of the United Nations; therefore, they have autonomy and make their own laws. It is important to know the in-country regulations, standards, and policies regarding trade, commerce, and the environment. The way business is conducted heavily influences the culture, traditions, and value systems. Government policies vary from one nation to the other. The adoption of policies and laws is mostly dependent on geographical location, country tradition and culture, and the influence of past colonial rulers. For example, in Latin America most of the countries adopt United States laws and policies, partly because of their trading and commerce relationship, whereas countries in Africa tend to derive their policies and laws from Europe.

Knowledge of in-country rules would be necessary protection in case of government interference or other discriminatory action that may thwart the implementation of project activities. In-country trade and environmental policies and overall governance provide the underpinning for business operations within the country's jurisdiction. Therefore, it is essential to understanding the trade and environmental policies and barriers within a country.

Various Bureaucracy and Governmental Functions and Structures. The various conditions, standards, and forms of governance dictate the functions and structures of a country. Nondemocratic countries exercise bureaucratic activities and functions that are obviously different from those of democratic governments. Religiously inclined governments, such as those in the Middle East, exhibit traditional (such as Arabic) forms of governance, which dictates the type and functions of their institutions.

An other obvious category of bureaucracy is dictated by the infrastructures and conditions within the country, such as the level of poverty, efficiency of public employees and personnel, inadequate equipment and technology, and the level of corruption within governmental institutions. These are important elements that a consultant should learn about before entering a foreign market.

Cultural and Traditional Customs and Practices. A very important aspect of doing business in a foreign country is to be aware of its customs, culture, and traditions. In most developing countries it is customary for people to respond to a foreigner's question and/or request by saying, "No problem." It is important to know that "no problem" does not mean that there is nothing to worry about; in most cases there is plenty to worry about. Cultural sensitivity is discussed in greater detail in Chapter 8. Different cultures and traditions cause communication difficulties and barriers. The consultant must be aware of the main language of communication and prepare either to study the language or to find an interpreter. Moreover, varying cultures and traditions may pose different communication problems. For example, most Africans talk with emotion and frequently use hand and finger gestures to make a point. Such nonverbal messages may be misinterpreted by Westerners as unprofessional. The Japanese bow their heads in salutation and if this gesture is not returned, it is implied that the other party is impolite and has no warmth.

Socioeconomic and Business Configurations and Forms of Transactions. Knowing the socioeconomic forms of business transactions in a country before commencing your consulting and business activities is very important. The Appendix XV provides basic economic indicators, including interest rates for each country. One has to understand the banking system to be able to receive and make payments. Some countries have foreign exchange markets that charge fees for exchange of currency. There are banks in many countries that offer services, such as money transfer, and issue checks, such as money orders. The business forms and configurations are very important, because in most developing countries transfer of money to and from foreign sources can be very difficult and frustrating. Government accounting systems and payment structures are unique from country to country; therefore, it is essential to inquire about these systems before entering another country to undertake any business activity. It is always important to check with the country's embassy or write to the appropriate ministry or department in charge of commerce. Appendix V provides information on each country's embassy in the United States, and information on each country's ministry of environment can be found in Appendix II.

Complex Forms of Registration and Licensing. The use of international standards is becoming popular globally, and, increasingly, many foreign contracts and Requests for Proposal (RFP) are requiring certification of individuals and consulting companies entering the international environmental market. The International Organization for Standards (ISO) has developed a series of standards that set a baseline performance standard applicable globally. The ISO is a nongovernmental body and whose series of environmental standards are developed by a team of industrial, nonprofit organizations, private volunteers, and standards organizations throughout the world. The ISO was formed to establish uniformity in standards throughout the European community as the unification of the European community approaches and reaches fruition. It also provides a vehicle to avoid international trade barriers that result in major conflicts over environmental standards that currently exist in certain countries. There is a series of standards under the ISO; the most important and significant applicable to practicing environmental management and consulting is ISO 14000. The ISO provides standards that a person or a company engaged in environmental management must satisfy to be certified for quality and ability to deliver and practice environmental consulting. The idea is spreading, and many more international environmental contract bids and RFPs will be requiring contractors to have this certification. Some of the ISO series, such as ISO 14000, include standards for:

- Life cycle analysis
- Environmental labeling and marking
- Environmental assessments and auditing
- Environmental management systems

Registration with ISO gives you an edge over competitors and provides you and/or your company with the accreditation of authority in your field of expertise. Most countries are greatly impressed with certificates and diplomas. They represent acceptance, respect, and authority. In some sense, they may even carry more weight than other qualifications such as experience and ability.

Licensing is another of the factors to be considered, because in some countries you are required to have a license to practice. For example, in the United States you may have to have a Practicing Engineer (PE) license—especially where design projects are involved—and to have the most basic practicing license, which is the state engineering license. There are other certifications in different environmental disciplines offered by professional organizations in the United States and in European countries. It is always advantageous to obtain a certification and/or a license. As stated earlier in this chapter, there is no requirement in the international marketplace to obtain a license to practice international environmental consulting. In respect to a basic engineering license, licensing requirements vary from one country to another, but in most cases general environmental consultancy work does not require such licensing.

Effective Networking and Creating Visibility. There is an old saying in the business world: "It is not what you know, it's who you know." This saying holds true in many ways, and we probably think about it in any business venture. It is interesting to note the importance of "Who you know" in developing countries, where it is necessary in all quarters of doing business. It is a must for "first time" international consultants before entering any transaction and/or seeking opportunities for contracts, especially with the government, to have significant contact with insiders. Without such networking, it is very difficult to achieve steady progress in your business dealings.

Networking is a way of developing contacts with a person or a company in the same or similar professional career or business. Networking provides an avenue for learning more about the professional trade, such as different techniques, new ideas, and experiences of others. It also provides a way of advertising yourself or your company and developing public relations. One of the keys to successful international environmental practice is self-marketing and getting to avail yourself to the multinational corporations, international financial institutions, the international donor community, and governments, to mention just a few. The idea is to make yourself known to your potential clients. International consultants are usually awarded contracts with clients through referrals or personal contacts. Therefore, developing brochures and pamphlets and placing advertisements are not necessarily the most effective ways of creating opportunities in the international consulting marketplace. There are many ways of networking and creating visibility, as shown in Fig. 3.3.

Advertising and Market Promotion. There are two ways to tackle advertising and promotion. One is to hire a professional advertising agency or consultant; the other is to do it yourself. The following are strategies for the "do-it-yourself" approach:

- *Develop pamphlets and brochures.* Pamphlets and brochures help to customize the key aspects of your consulting practice or your company for presentation to potential markets or clients. You must develop a simple, easy-to-read pamphlet, highlighting your strengths and capacities. You must specifically indicate areas of your services and expertise. It is advantageous to develop more than one pamphlet, each addressing a specific audience or target market.
- *Develop a mailing list and establish direct mailing.* Creating a clientele mailing list is the first step in creating a market to sell your services. You can develop the list from advertisers in international environmental journals and magazines. There are in many cases available lists of attendees at major conferences and seminars. Obtain lists of attendees to major international functions inclusive of professional environmental confer-

- Join international professional organizations.
- Attend international environmental seminars and conferences.
- Participate in international donor organizations' public activities.
- Establish and maintain contacts with an informal circle of mentors and advisors.
- Pay regular visits to foreign countries in which you are interested in doing business.
- Identify other international consultants and develop collaboration opportunities.
- Develop, collect, and maintain a contact list, an address book and subscribe to international environmental journals and international trade magazines.

Figure 3.3 Ways of Networking and Creating Visibility for a Successful International Environmental Consulting Practice

ences and nonprofessional functions which may or may not relate to environmental activities but would be useful for international sources. Obtain directories of environmental professionals, attorneys, and business executives. Collect business cards from potential clients and those who are in positions of influence to other target markets or potential clients.

- *Make presentations at international events and professional conferences.* One way of developing your credibility is through presentations before your potential clients and colleagues. Technical papers and presentations can establish you or your company as an authority on a particular subject. This is a strategy you can use to market yourself and strengthen your position in gaining major contracts.
- *Exhibit at international conferences and trade fairs.* For manufacturers of environmental technologies, trade shows offer a way of demonstrating products to a wide range of potential clients. If you are an individual consultant, you can display examples of your work performed or specific projects that you have completed. This helps potential clients needing similar services to remember you or your company.
- *List your company or practice in directories, catalogs, or other international publications.* Telephone books, business journals, directories, news magazines, and newspapers are sources of reference for most clients who are searching for consultants to undertake a specific project with specific expertise.
- *Write articles.* Developing and writing professional articles within your area of expertise and publishing in journals will provide exposure to potential clients and opportunities in foreign markets.
- *Use radio, TV, and billboards.* Place advertisements on radio, television, and billboards in countries where you prefer to conduct business and/or consultancy.

3.3.3 Selection of Optimal Alternatives

The framework for effective evaluation and selection of alternatives should be conducted by selecting a tool that is in line with the values, culture, and the mission setforth by the company. Some of the most important tools often used are "weighed" approach or a comparative assessment in combination with cost benefit analysis, cost effective analysis, simulative analysis, contingent valuations, input-output analysis, macroeconomic models, and decision-making theories. Some common errors in choosing a tool in the selection process are:

- Selecting a tool that does not fit in the core values of the organization
- Selecting a tool that would provide a desired decision for the optimal plan instead of an honest analysis

- Selecting a tool that has been used previously and did not provide the best selection of a plan

The strength of this approach of careful analysis is that it provides a broad view of all the variables. The author has always pointed out to decision-makers that solely weighing heavily on economic value as the main criterion of the selection can be fatal because there are very dynamic variables when one is dealing with foreign countries. These dynamics may be unknown to the decision-maker.

Finally, peer review is always a good approach to getting a second opinion from an outside source. Notwithstanding the fact that the ultimate and final decision rests with the stakeholders, that is the company and its directors.

3.3.4 Implementation of Action Plans and Alternatives

This is the final stage in the planning process: the phase of implementation of selected optimal plan. An Action Plan must be developed and implemented. The plan must include time-tables and timeliness for execution of planned activities. Most important is a clear task and assignment to all team members. Clear roles and responsibilities should be developed and understood by each person on the team implementing the plan.

4

INTERNATIONAL PARTNERSHIPS AND JOINT VENTURES

The concepts of partnership and joint ventures, although similar in theory, exhibit many differences in practical application. Both are rooted in the fundamental principle that this book defines as the "commonsense" approach to effective utilization of resources and capabilities of the parties involved to pursue similar interests and reach a common goal. International environmental consulting practice and activities become easier, less stressful, and less cumbersome when there is a partner stationed in the country where you intend to practice. Success in most cases is derived from:

- The ground presence (direct or indirect) of the individual consultant. Direct ground presence means physical presence, direct networking, and marketing with partners, government, and private and public sectors.
- The ground presence (direct or indirect) of the corporation or company
- The ground presence (direct or indirect) of the manufacturer of environmental products,
- The technical, financial, and political strengths of partner

These are factors that, when combined with a partner in-country, can result in success for a short-term project or in the long-term where there is potential for future growth in market opportunities.

The principles of partnering and joint venturing have evolved, and one can classify this movement as the "teamwork phenomenon" that is reflected in the current wave of corporate attitude change, corporate reengineering, and cultural change. As we grow into a globalized economy and enter global trade and business competition, corporations and companies are all searching for—and finding—ways to be efficient, cost-effective, and productive and to offer optimum customer satisfaction. Therefore, partnerships, joint ventures, and mergers are evolving as the restructuring tools of many corporations to achieve shared goals and be competitive in the global market. In the past, the traditional handshake of acknowledgment of a partnership agreement was an adequate formality. Today, partnerships have become more sophisticated, providing broader involvement and gradual and systematic construction of relationships between parties. Partnership involves two or more persons, entities, or stakeholders pursuing a mutual, strategically structured commitment to joint administration of projects and activities with a view to making profits.

4.1 THE PARTNERING PROCESS

The framework of partnering is intended to strengthen and complement all stakeholders' capabilities in a particular business project or toward overall business goals and objectives. It is essential, therefore, to evaluate the merits and demerits of a potential partner in regard to attributes and elements such as:

- Identification and evaluation of goals and assessment of common interest
- Identification and evaluation of debt, equity, and assets
- Identification of potential lines of communication and evaluation of the degree of commitment and level of cooperation in problem solving
- Identification and evaluation of risk

The rationale for establishing a partnership is to create a working relationship based on mutual understanding, trust, and share commitment to shared resources among the stakeholders (Figure 4.1a, b). Evaluation of the basic principles of partnership should be performed well in advance of actually signing a partnership agreement. The following are basic components that require careful evaluation:

- Level of trust toward the other partner
- Extent to which stakeholders have mutual goals and objectives for a specific project or general business
- Level of equity to ascertain the degree of interest by equity partners to commit to the project development and mission.

The final category of importance is the cost-benefit analysis of the partnership. Factors to be evaluated in the analysis may include, but are not limited to:

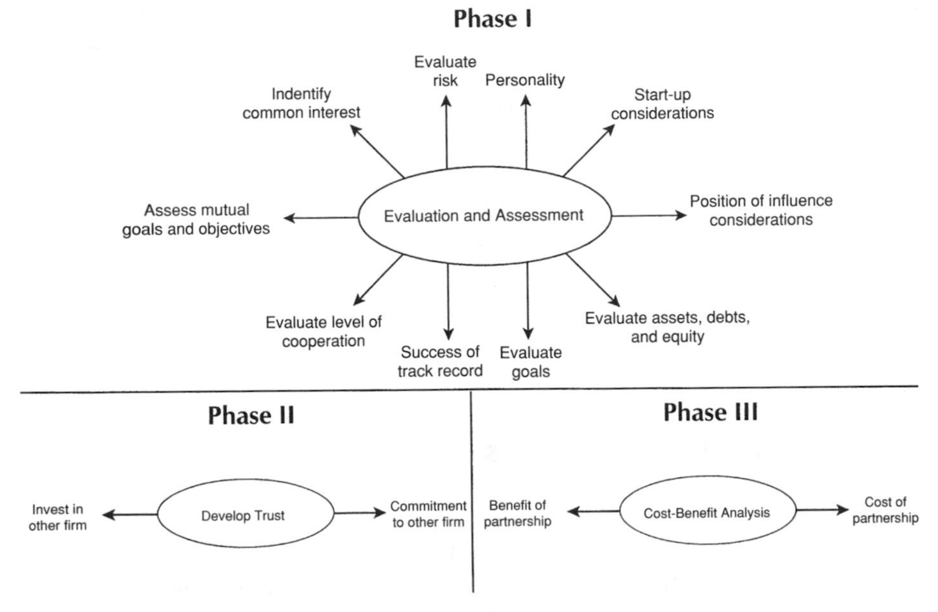

Figure 4.1a Evaluation and assessment for partnering (*Source:* Peter A. Sam, 1997)

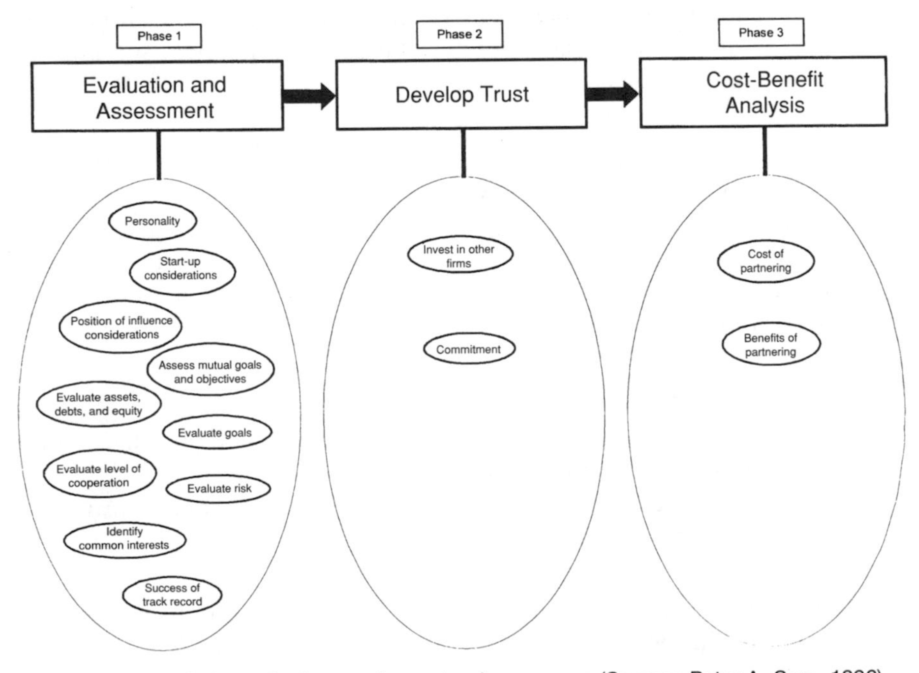

Figure 4.1b Schematic diagram for partnering process (*Source:* Peter A. Sam, 1996)

1. *Start-up Cost.* In cases where you find yourself or your company financially stressed in staring a consulting practice, the decision to choose the alternative, partnership, should be based on whether such partnering improves the ability to acquire start-up capital and decrease the time needed to start operations. A partner could be an equity investor who will provide needed capital for start-up cost.

2. *Position of Influence.* Having partners and contacts in-country who are in a position to have socioeconomic and political connections can provide easy and workable avenues for business connections and transactions in that country.

3. *Physical Presence.* Having a physical presence in the overseas location rather than having to temporarily reassign staff to the project site presents opportunities for cost savings. It is very expensive to relocate staff or a subcontractor to a country overseas to implement projects or activities associated with an international environmental practice. The per diem rates and other ancillary overseas expenditures are quite high, and it is advantageous to have a local in-country presence or a partnership for the overseas project sites. Physical presence also provides an advantage over competitors in that you or your company will have globally neutral characteristics, as well as visibility, which reduces the level of effort and expenditure in advertising and marketing. Expenses for travel, education, and training in areas other than product knowledge can be avoided; the in-country partner will not require such training owing to his or her nativity in and familiarity with the region.

4. *Marketing and Sales.* Savings in marketing and sales costs can also be realized through having in-country partners. The differences in cultures and sensitivities in the community in which you are planning to do business can pose a difficult barrier to overcome. The cost of hiring a consultant and implementing strategies can be avoided if your partners are natives and reside in the overseas location of your business. You should always ask yourself: "What are the least painful, least expensive ways to present information to a global audience within the international market?" The best and most cost-effect approach is to have local native partners (agents) to market and sell your services. A physical presence, as mentioned earlier, aids in marketing your services, selling your products, and enhancing prospects for new clients.

5. *Human Resources Capacity.* Human resources capacity includes a cost assessment of your prospective partner, internal human resources, and the benefits of administrative procedures. The benefits and strengths of a partner's special skills, such as sales and marketing skills and other expertise that you or your company do not have, can result in immense cost savings because you do not have to develop new expertise or hire and train new staff.

4.1.1 Benefits of Partnering

Partnering can create excellent opportunities for accelerated business growth as well as an efficient and effective international environmental consulting practice. Partnership usually provides an advantage in seeking financial loans to commence a consulting practice abroad. The notion of the combined financial resources can strengthen the growth potential of the business simply because there is a sufficient financial base between the two partners for overhead expenses and available running capital. Finally, unlike joint ventureship, partnership does not normally require legal and extensive bureaucratic formalities. One should evaluate on a case-by-case basis the need to consult an attorney to prepare a Consent Agreement to avoid future misunderstanding and confusions and to secure maximum protection for both parties involved in the partnership. Legal obligations such as taxation and personal responsibilities rest on each individual partner. Partners often negotiate how to address and treat taxes. This is because in a partnership each partner is responsible and liable for taxes. Each partner carefully examines how the other partner will calculate taxable income. Together, the partners usually settles how taxable income will be handled. Year-end tax responsibilities are borne by each partner. The financial statements and calculation process in this case provides the possibility of tax loopholes of which partners can take advantage. There are tax-planning techniques frequently used by partnerships to take advantage of tax rules in establishing the partnership in the taxation years of the partners, thus achieving a deferral in income reporting.

Other advantages can be achieved through the following:

- Special tax incentives, tax benefits, or avenues for tax loop-holes
- Export credits requirements
- Legal protection for trademarks and patents
- Technology transfer policies

4.2 THE JOINT VENTURE PROCESS

A joint venture is a form of partnership based on the mission and objectives of the venture and the reasons for which the ventureship is to be established. One interesting difference between ventureship and partnership is that partnerships can be ventures but ventures may not be partnerships. A joint venture is a form of relationship in which personalities and internal commonalities are not considered or merged into the objectives all stakeholders are pursuing. The objectives are always set in stone—well defined and very specific. Usually, the main objective is to join capital and expertise to attain a commercial goal. In this manner both venturers preserve capital, contribute expertise, and spread risk. Joint ventures can be classified in approximately six categories:

- Limited partnerships
- Unlimited liability partnerships

- Limited liability partnerships
- Corporate joint ventures
- Single-venture partnerships
- Consortia

Joint ventures are usually created from available legal and tax-related vehicles for a particular activity within the international market. The most notable advantage in ventureship is often the tax break consideration. In certain instances some countries have policies that require a foreign company or an individual entity to acquire a local or an indigenous partner for a certain specific commercial sector activity. The United Kingdom, for example, requires joint ventureship among consultants, manufacturers, service and supply companies, and contractors (persons and entities) entering the petroleum industry operation in the North Sea. In some developing countries, such as Ghana, government policy on their divestiture programs to sell nationally owned corporations and industries requires the foreign investor to involve, at the maximum extent possible, local and indigenous partners.

4.2.1 Selection Process

The intention to consider ventureship, by its very nature, should be used as the premise and a criterion for selection of a venture partner (Figure 4.1b). The other factors to evaluate in the process of considering ventureship are, in part, to:

- Evaluate the strengths and weaknesses of the potential partner.
- Evaluate the expertise, skills, and qualifications of the potential partner.
- Evaluate the credibility, acceptability, and registration of the partner within the specific jurisdiction of your ventureship.
- Evaluate the tax position of the potential partner, tax criteria, and tax laws that are of advantage or disadvantage to the potential partner.
- Evaluate the impact of antitrust and monopoly implications and the effects of joint venturing with the potential partner.
- Evaluate and assess the potential partner's suitability to the joint ventureship.
- Evaluate the financial and human resource capacity and strengths of the potential partner.
- Evaluate the research, development, and technology capacity of the potential partner.
- Evaluate financial and insurance rating of potential partner's company.
- Evaluate the level of conflict caused by the potential partner's corporate structures and business interactions and activities.

- Evaluate and assess communication barriers.
- Evaluate and assess legal and tax implications.

4.2.2 Benefits of Joint Venturing

Many countries have policies requiring joint ventureship for foreign companies. Nevertheless, joint ventureship provides several advantages apart from in-country government acceptance. Capital contribution by a foreign partner aids the local partner in reducing the foreign partner's overall risk exposure, and the local partner benefits also from acquiring capital. The local partner contributes some knowledge of the in-country business environment and other in-country logistics.

Tax laws in one country many benefit the international venturer or the domestic venturer, particularly when the structure or formation of the ventureship is based on a double-taxation agreement. Joining of capital and expertise by small practitioners and companies has frequently resulted in real benefits to the venturers, such as providing the ventureship with a greater share of the market and more opportunities to compete with rivals. This provides the venturers a feasible method of achieving a business goal. One should consult a tax attorney, an accountant, or an expert in international joint ventureship to assess the optimum benefits of a prospective venture. In joint venturing, benefits can be derived from the following:

- Special tax incentives, tax benefits, or avenues for tax loop-holes
- Export credits requirements
- Legal protection for trademarks and patents
- Technology transfer policies

5

PROPOSALS, BIDDING, CONTRACTS, AND NEGOTIATING

This chapter describes the process of winning a project contract: developing a proposal, the bidding process, developing a contract, and negotiating terms (Figure 5.1). Each topic is discussed in terms of the major elements and applicable issues that should be considered by an individual planning to enter international consulting practice.

5.1 PROPOSAL DEVELOPMENT

Developing proposals is an essential and critical aspect of the consultant's Scope of Work (SOW) to perform a task or project. A technically sound, cost-effective SOW is the cornerstone of what the client will use to assess whether the proposal meets its needs and, furthermore, addresses the technical problems that must be resolved. The SOW becomes a guiding light in designing the Work Plan (WP) to implement the project. Proposals also forge an understanding between the consultant and the client in regard to the task or project to be performed and the technical solutions to the problem stated.

Clients are inclined to seek specific qualities that match the needs of specific projects/tasks. The selection of consultants and award of contracts depend on how well the proposal is written and its technical merits. A basic proposal contains an introduction, an Issue/Problem Statement, a Scope of Work, Cost and Pricing Factors, Price and Timeline for Delivery, and Payment Terms. Other contractural provisions and technical approaches are included in most cases. A good proposal addresses major elements that clients would

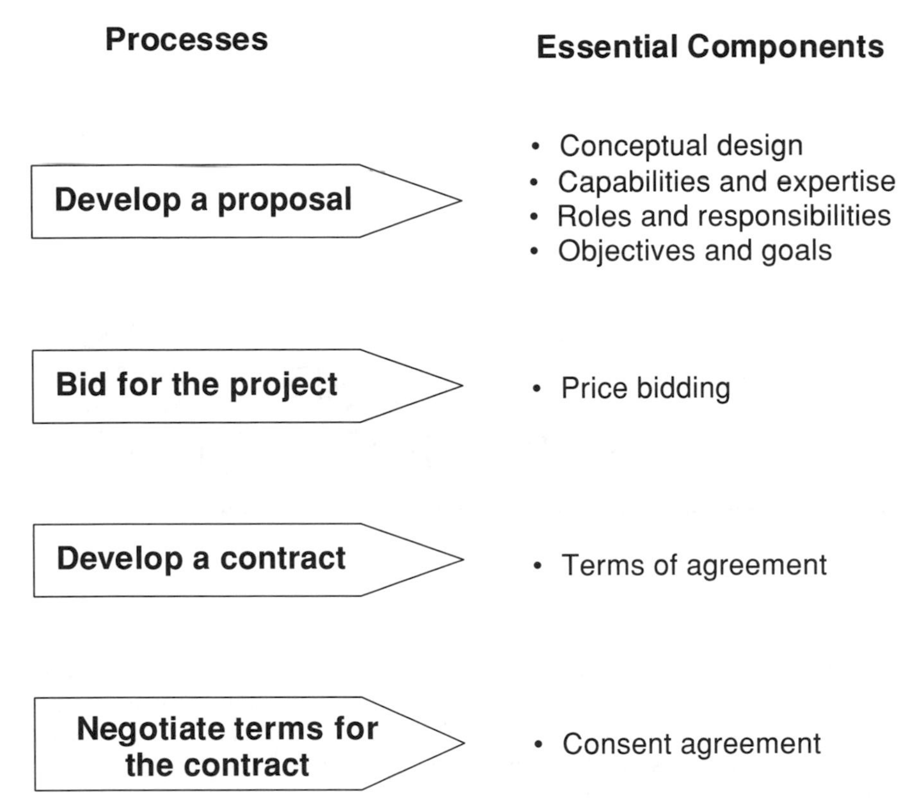

Figure 5.1 Process for winning a project contract award

look for in judging the strengths and merits of the proposal. Attention to the following details can strengthen a proposal:

- Sound technical approaches
- Conformity to the clients' technical requirements
- Level and degree of expertise
- Skills and capacity to deliver services
- Knowledge (current and past experience relating to project-specific problem)
- Compatibility: consultant's characteristics, attributes, and potential working relations with clients
- Cost-benefit approach demonstration/cost savings approach
- "Best deals for the buck" demonstration

As a consultant, your initial strengths will rest in the quality of your proposals. It is through the packaging of the proposal that you get a chance at winning a contract. Therefore, you should satisfy the client's needs by addressing at least the aforementioned qualities to make the proposal palatable to the client. In many cases the client may specify criteria and instructions for the preparation of a proposal. These must be met, because the client will use the criteria as an evaluation tool to rate each proposal.

5.1.1 Types of Proposals

The marketing of your practice is a "leap" to proposal development; a proposal is an intermediary to selling your services and consultancy. Whenever a marketing campaign is launched for your consulting practice, you can expect one of two things: rejection by the client as an unsuitable consultant for its needs and wants or impression with the expertise and services you offer. There are various types of proposals; which is used depends on the client's wants. The following are a few examples:

- Billing and fees (Price Quotation Proposal)
- Expertise, experience, and skills (Qualification Proposal)
- Statement of work (Conceptual Proposal)
- Procurement and Delivery (Sales and Services Proposal)

A proposal can be developed (1) in response to a Request for Proposals (RFP) or (2) as an unsolicited proposal. These are the major avenues for generating international consulting business for your firm. It is no easy task, but the advantage lies in your ability to develop an excellent response to an RFP. The challenges and the competition will be great, because your competitors will be targeting the same sources that are discussed in this chapter. Generally, leads to the public RFPs can be found in trade publications and existing market data. Utilization of these major sources will assist you in:

- Winning significant contracts
- Developing proposals
- Targeting new markets and territories
- Keeping abreast of new developments and surveying your competition

Response to Request for Proposals (*RFPs*). RFPs are initiated by both governmental and nongovernmental entities for the public to bid on a specific project. Within the international arena, there are several sources of RFPs. The major sources of governmental entities issuing RFPs are, in part:

- The international donor community
- International development agencies

- In-country governments
- The governments of the United States and other developed nations

There are, however, fewer RFPs generated by nongovernmental sectors in comparison with the number issued by governmental sectors within the international environmental consulting arena, particularly in developing countries. The major nongovernmental sectors issuing RFPs are:

- In-Country companies and businesses
- Multinational corporations

Unsolicited Proposals. These are proposals that come out of your marketing and sales efforts. Most of them result from referrals by former clients, friends, and acquaintances.

5.1.2 HOW DO YOU OBTAIN RFPs AND CONSULTING JOBS?

There are several sources that can be used to obtain RFPs:

- Internet access
- Monthly publications
- Direct contact
- Registration as a subcontractor
- Registration with international donor organizations and developing agencies

Internet Access. The web sites and gophers established by international development agencies, international organizations, and international financial institutions can be accessed for information and data. The development business provides SCAN-A-BID, which is an on-line database that can be accessed via personal computer and a modem.

Sources of Information on the Internet. Several governmental agencies and private organizations provide vast amounts of information. The major web sites include the following:

The USAID Web Site
Http://www.info.usaid

Commerce Department Web Site
HTTP://www/ucg.com/nifp/nifp.html

Federal Register Web Site
HTTP:/cos.gdb.org/repos/fr/fr-intro.html

Multilateral Development Bank Web Site
www. ita.doc.gov//mdbo

National Trade Data Bank Web Site
Http://www.stat-usa.gov

United Nations Home Page Web Site
Http//www.undp.org

Business Publications Home Page Web Site
Gopher://gopher.enews.com:/11/magazines/category/business/international

Monthly Publications. There are a number of publications from which you can obtain RFPs. The United States government publishes the *Federal Register,* and some of its agencies have individual regular publications on RFPs. The *Commerce Business Daily,* like the *Federal Register,* includes monthly information about RFPs. These publications provide quick information and data about consulting opportunities opening up in every continent in the world. Many other countries prepare similar publications, which in many instances provide information on the status of international financial institutions' projects in the pipeline, from a project's inception, by country and by sector, through the bidding process, up to and including information on who awards and wins a contract.

Other governments utilize local newspapers and magazines for these announcements, among which are the following:

- *Wall Street Business Journal*
- *West Africa Magazine*
- *Business Week*
- *The Economist*
- *Export Today*
- *Far East Economic Review*
- *Eastern Europe Business Bulletin*
- *Engineering News Record*

Direct Contact. Direct contact means contacting specific organizations, potential companies or corporations, or the appropriate government agency in-country, to find out about any pending RFPs. You may also contact the embassies in Washington, D.C., to inquire about any RFPs that are pending. The following are addresses of international organizations:

The World Bank
Development Business Liaison Unit
Room E-8050
1818 H Street N.W.
Washington, DC 20433
Phone: (1)(202)458-2397

The United Nations
Development Business
P.O. Box 5850, Grand Central Station
New York, NY 10163-5850
Phone: (1)(212)963-1515

The Commerce Business Daily
U.S. Department of Commerce
14th Street and Constitution N.W.
Washington, DC 20230-001
Phone: (1)(202)377-4878

Federal Register
National Archives
Records Administration
Washington, DC 20408-001
Phone: (1)(202)523-5230

United States Agency for International Development (USAID)
320 21st Street, N.W.
Washington, DC 20523
Phone: (1)(202)647-4000

Registration as a Subcontractor. Big companies engaged in international business may keep prospective subcontractors in their databanks. It is very useful to contact many large companies to provide them with your profile, thereby making yourself available as a registered subcontractor.

Registration with International Donor Organizations and Developing Agencies. Many international donor and developing organizations provide registration opportunities for consultants and firms. You must identify those organizations you wish to contract and register with as a consultant. Usually, upon acceptance of your request to register as a consultant, your profile is placed in a databank and reviewed against any projects that call for your firm's expertise and capabilities. Appendixes III and IV of this book list many of these organizations and their addresses for quick reference.

5.2 CONTRACTS

A contract may generally be defined as a mutually binding legal relationship obligating a seller to furnish supplies or services, and a buyer to pay for them.

Specifically, a contract is an agreement between a client and a contractor/consultant that spells out the services, work performance, and products to be provided by the contractor/consultant. A contract sets forth functions relating to the project from the moment it is awarded until final payment has been made and the contract is completed. There are various types of contracts, some of which are described in Table 5.1.

Table 5.1 Major Types of Contracts—Merits and Demerits

Contract	Features	Applicability	Advantages to Contractor	Disadvantages to Contractor
Firm Fixed Price	Client pays fixed price (established before award), which is not subject to any adjustment regardless of contractor's cost experience.	Used when: 1. There are reasonably definite design or performance specifications and 2. A fair and reasonable price can be established at the outset.	1. Potential for higher profit. 2. Minimum control by client. 3. Fewer administrative costs.	1. Greater assumption of financial and technical risks. 2. Vigilance required to initiate and substantiate change claims.
Cost-Plus-Fixed-Fee	Client pays allowable cost plus a negotiated, fixed fee (profit). Fixed fee does not vary with actual costs, but may be adjusted for changes in work to be performed. May be completion or term form.	Used where performance is uncertain and accurate costs estimates are impossible.	1. Low risk. 2. Risk of loss of client's property transferred.	1. Maximum control by client. 2. Lower fees because of lower risks.
Cost-Plus-Award-Fee	Client pays allowable costs plus base fee (does not vary with performance) and all or part of an award fee (based on subjective government evaluation of contractor's performance). Evaluation and payments of award fee made periodically (usually every 3–4 months) during performance.	Used where a cost reimbursement contract is appropriate and it is important that contractor be provided motivation for excellence in contract performance in areas such as management, quality, timeliness, ingenuity, and cost effectiveness.	1. Possibility of increased fee. 2. Reward for good management and good performance. 3. Limited risk.	1. Limit on fee (usually 10%) and/or the % of the contract amount. 2. Complexity of negotiations. 3. Burden to "prove" that award is justified. 4. Performance may be affected by client monitoring/technical direction.

(*continued on next page*)

Table 5.1 (*Continued*)

Contract	Features	Applicability	Advantages to Contractor	Disadvantages to Contractor
Labor-Hours and Time-and-Materials	Client pays a fixed rate for each hour of direct labor worked by contractor, up to a negotiated ceiling on the total price. (Time-and-materials contracts also provide for payments for materials at cost). Indirect costs and profits are all incorporated in the fixed hourly rates.	Used typically for engineering and design services, repair, maintenance or overhaul work, or in emergency situations. Least preferred contract type. Must have appropriate government surveillance during performance.	1. Potential to maximize profits. 2. Minimal risk.	1. Constant client surveillance.
Indefinite Delivery/ Indefinite Quantity	Contract is somewhat open-ended, i.e., does not specify delivery or performance terms. Payments may be on a fixed price, cost reimbursable, fixed rate per item or service, or a labor-hour basis. Orders are placed against the contract after award. Indefinite quantity contracts provide for a "minimum" and a "maximum" quantity.	Used when the exact time and/or place of delivery is not known at the time of contracting. May be used for either supplies or services.	1. Guaranteed minimum quantity (unless client terminates contract). 2. Potential to maximize profits, depending on payment provisions.	1. No control over scheduling orders. 2. No guarantee of orders beyond contract minimum. 3. Possible substantial amount of client's surveillance, depending on payment provisions. 4. Requires high degree of management involvement.

Peter A. Sam, 1996. Modified from U.S. EPA Contract Administration Training Manual, PCMD, 4/89.

There is no set format for developing a contract with a client, but you must always address and include the following provisions:

- Parties to the agreement
- Dispute resolution
- Fees
- Terms of payment
- Work assignments (services, products, procurement) and schedules for deliverables
- Monitoring and oversight obligations
- Approval and concurrence process
- General obligations of both parties
- Force majeure and excusable delay (covering acts of God, unavoidable delays, unenforceable clauses, indemnification, warranties, and assignments of rights and responsibilities)
- Signatures of authorized representatives of both parties
- Detailed work assignment and effective date of agreement
- Termination date of agreement
- Statement of commitment of client to retain the services of the consultant or firm for the specific project
- Utilization of previously obtained data from previous contractor
- Covenant not to sue
- Termination and satisfaction

5.3 BIDDING

Bidding refers to the process of providing specific statements of the work and services to be rendered in a proposal. Clients prefer to set a procedure for bidding. Most procedures require statements of qualification and expertise, scope and design, and other specific components that the client may need to make his or her selection. This procedure is an effective way for a client to evaluate a bidder's experience, skills, and expertise against price. Ultimately, the client will select the consultant or firm presenting the best design, skills, experience, and expertise with lower cost.

It is critical to demonstrate in your bid proposal that you have applied the maximum cost-savings approach. You should use the most effective and efficient cost factoring application in deriving your pricing. In some cases the client will provide the pricing structure for the work, labor rates, reimbursable local and foreign cost, and so forth. You should obtain the latest exchange rates for the country in question by contacting the in-country central bank to calculate the conversion of the U.S. dollar, or your country's currency, to the local currency for the project. Finally, your bid proposal must be broken down in the format provided by the client's RFP or the pricing score sheet. Failure to follow the format usually results in rejection of your proposal.

5.4 NEGOTIATING

Negotiating simply means persuading others to agree. In the international market, negotiations take a lot of work and include unique factors that are not usually considered when negotiating in your home country—cultural norms, sex stereotypes, and traditions. Various techniques, theories, strategies, and gambits can be used in negotiations. When you are in a foreign country, however, no matter how well you have mastered these techniques and theories, the playing field changes and the best technique is to be very sensitive to the country's culture and traditions.

It is a must to prepare before entering into negotiations in a different country. The key to successful negotiating is strategic planning and doing your homework. The following are guidelines for your prenegotiations planning:

- Know as much as possible about your opponent/client and the essential elements of the negotiations.
- Use in-country contacts to obtain information about the opponent/client. If you do not have contacts in-country, get information from the country's embassy, trade journals, in-country government agencies, in-country chambers of commerce, etc.
- Be aware of the cultures, traditions, and taboos of the country and its people. Remember to start gathering information and doing your research early. An example of a culture norm in developing countries, and even in most European countries, is that most people are identified by their titles and credentials, such as Dr. ABC, Director of XYZ. Developing countries are quite particular to educational achievements and credentials, and their people are easily offended if not addressed appropriately. Therefore, it is important to remember the names and credentials of the opponent's negotiating team.
- Try to anticipate the needs and wants of the client/opponent.
- Prepare documentation and proof to back up any issue or position that your opponent/client may be suspicious of. In many developing countries, there is a saying that "seeing is believing." This is a way to develop trust between you and your client/opponent.
- Sort out the major issues and develop alternative solutions that play in your favor.
- Develop and negotiate your strategic and contingency plans.
- Assemble your team and assign tasks and roles. Assign a leader or chief negotiator, a note taker, and an observer.

6

INTERNATIONAL CONSULTING MARKET OPPORTUNITIES AND PROSPECTS

This chapter examines the global opportunities for international environmental consulting. A brief overview of the environmental problems and issues in each continent is provided, in addition to a discussion of lending and financing sources for environmental activities that increase the prospects for environmental consultants.

There are many obstacles associated with entering and conducting environmental consulting business overseas. But with reduced opportunities in the United States, domestic environmental market, which is undergoing fierce competition, increasing consolidations and mergers, and bankruptcies of many small environmental companies, opportunities in the international market will yield a higher return than investment in the domestic market. The United States' environmental market is forecast to grow at an average rate of between 4 and 5 percent, outpaced by the average growth of the environmental market in the rest of the world, which is forecast to be more than 10 percent (Environmental Business International Inc., 1995). The environmental markets in foreign countries, especially in developing countries in eastern and central Europe, Asia, Latin America and Africa, are booming—with virtually no competition. In Africa for example, the World Bank in 1998 under its Environmental Project Portfolio, a total project cost approximately $1.3 billion. In this situation, profit margins can be lucrative because pricing competition is absent. These countries are catching up with the global agenda for change and are on the developing edge of environmental protection, restoration, and management. In comparison to the United States and other industrialized nations in the West, the developing countries are developing their environ-

Table 6.1a Growth in the Environmental Market 2000

	1994	avg.	1995	1996	1997	1998	1999	2000	2001
United States	165.5	4%	172.1	179.0	186.2	193.6	201.4	209.4	217.8
Canada	10.8	5%	11.3	11.9	12.5	13.1	13.8	14.5	15.2
Latin America	6.6	12%	7.4	8.3	9.3	10.4	11.6	13.0	14.6
Western Europe	127.4	4%	132.5	137.8	143.3	149.0	155.0	161.2	167.6
East Europe/Russia	6.4	8%	6.9	7.5	8.1	8.7	9.4	10.2	11.0
Japan	65.3	4%	67.8	70.4	73.1	75.8	78.7	81.7	84.8
Rest of Asia	14.2	17%	16.6	19.4	22.7	26.6	31.1	36.4	42.6
Australia/New Zealand	6.2	5%	6.5	6.8	7.2	7.5	7.9	8.3	8.7
Middle East	3.8	6%	4.0	4.3	4.5	4.8	5.1	5.4	5.7
Africa	1.8	10%	2.0	2.2	2.4	2.6	2.9	3.2	3.5
	408	**5%**	**427**	**448**	**469**	**492**	**517**	**543**	**572**

Source: Environmental Business Journal, vol. VIII, no. 8, 1995. San Diego, CA, units in $ billions.

mental management and building capacities at a steady rate, which is leading the development of the environmental markets in the region. In contrast to some of the Western industrialized nations where many environmental sectors are exhibiting market stagnation, the developing nations are showing signs of a growing prosperity and increased market size. *Environmental Business International* reported in 1995 that the global environmental market has reached a comfortable level at $408 billion for 1994 and projected to reach $543 billion in 2000 as shown in Table 6.1a. Figure 6.1 presents major environmental sectors in global regions and their respective shares of the global environmental market. The most promising segments of the market are solid waste management, pollution control, waste minimization, water quality, and

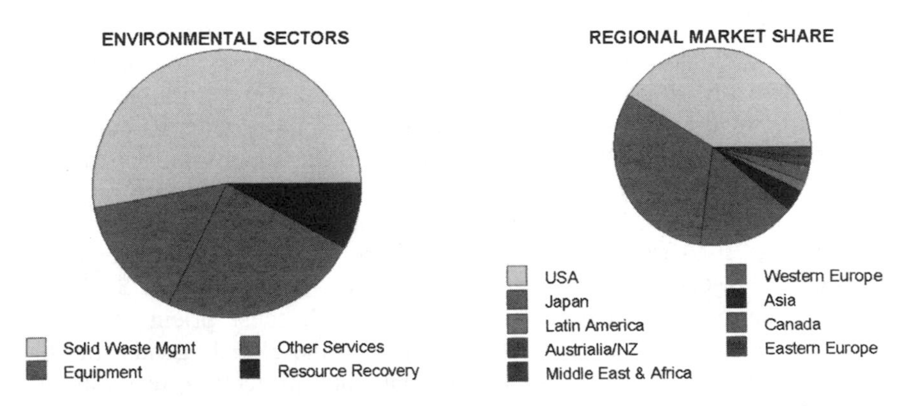

Figure 6.1a The $408 billion global environmental market (*Source: EBJournal,* 1995, Vol. 8, Environmental Business International Inc., San Diego, California)

	United States	Western Europe	Japan	Asia	Latin America	Canada	Australia/ New Zealand	Eastern Europe	Middle East	Africa	Total $ (billions)	Total %
Equipment												
Water Equipment and Chemicals	13.2	10.0	4.8	2.0	1.0	1.1	0.6	0.7	0.4	0.3	34.	8.3%
Air Pollution Control	11.6	7.0	4.4	0.7	0.3	0.5	0.2	0.4	0.3	0.0	25.6	6.3%
Instruments	1.9	1.5	0.7	0.1	0.1	0.1	0.1	0.1	0.1	0.0	4.6	1.1%
Waste Management Equipment	11.0	8.7	3.8	1.0	0.4	0.8	0.4	0.4	0.2	0.1	26.7	6.5%
Process and Prevention Technology	0.8	0.5	0.5	0.1	0.0	0.1	0.0	0.0	0.0	0.0	2.0	0.5%
Services												
Solid Waste Management	31.0	28.1	20.2	2.6	1.2	2.1	1.2	1.0	0.7	0.3	88.3	21.6%
Hazardous Waste Management	6.4	5.0	3.6	0.4	0.3	0.4	0.2	0.3	0.2	0.0	16.7	4.1%
Consulting and Engineering	14.7	8.0	1.0	0.6	0.3	0.9	0.5	0.3	0.2	0.1	26.5	6.5%
Remediation/Industrial Serivce	8.5	3.5	1.1	0.3	0.1	0.4	0.2	0.2	0.3	0.0	14.7	3.6%
Analytical Services	1.6	1.0	0.0	0.1	0.1	0.1	0.1	0.1	0.0	0.0	3.1	0.8%
Water Treatment Works	25.8	20.8	8.8	2.0	0.8	1.8	1.1	0.6	0.3	0.1	62.1	15.2%
Resources												
Water Utilities	24.3	18.8	10.1	3.3	1.6	1.8	1.2	2.1	1.0	0.6	64.9	15.9%
Resource Recovery	13.1	13.0	5.8	0.9	0.3	0.6	0.3	0.3	0.1	0.1	34.6	8.5%
Environmental Energy	1.6	1.5	0.5	0.3	0.1	0.1	0.1	0.1	0.0	0.1	4.4	1.1%
TOTAL $ (billions)	165.5	127.4	65.3	14.2	6.6	10.8	6.2	6.4	3.8	1.8	408.0	
TOTAL %	40.6%	31.2%	16.0%	3.5%	1.6%	2.6%	1.5%	1.6%	0.9%	0.4%		

Figure 6.1b The global environmental market, 1994

sewage disposal/wastewater. The slow areas or segments are hazardous waste management and air quality.

Table 6.1b provides the EBJ's top 50 environmental markets in the world. Growth in the global environmental market is projected to increase steadily, as indicated in Table 6.1a.

The global political pressure and insistence for environmental management, coupled with the growing global consensus that economic development has to be complemented with sound environmental policies, provides, in part, a major driving force for future demand for international environmental consulting. The reduced global political tension resulting from the end of the Cold War has shifted attention to investment in sustainable development for the coming years. The United Nations Conference on Environment and Development (UNCED) in Rio de Janeiro provided the platform for increased environmental management and socioeconomic activities worldwide. In many nations, especially in the developing countries where environmental technologies and expertise are lacking, there are numerous opportunities for foreign consultants as local and central governments seek experts to implement environmental projects. The UNCED Agenda 21 allocated international financing and aid to assist countries in implementing programs for environmental management.

Multilateral lending institutions such as the World Bank and the International Monetary Fund have instituted environmental screening of the development projects they fund; therefore, more Environmental Impact Assessments (EIAs) and (EISs) will be required, which in turn generate opportunities for international environmental consulting in developing countries.

Prospects for international consulting within the driving forces of environmental management were discussed in Chapters 1 and 2. This chapter provides information on potential environmental problems and needs in each continent and, in some cases each country. This will alert the reader to future consulting market opportunities that may arise as a result of environmental problems within a particular continent or region.

6.1 CONSULTING OPPORTUNITIES

Consulting opportunities are created by in-country priorities and agendas for the environment, natural resource management, and protection of human health. On the other hand, opportunities for environmental consulting can be created by the financial and technical support from many outside sources. The purpose here is to provide the reader with an overview of the environmental issues and priorities of each continent, as well as a brief description of major sources where environmental consulting opportunities may arise. These major sources are:

- International development assistance agencies

TABLE 6.1b EBJ's Top 50 Environmental Markets in the World

Rank	Country	Market Size	GDP	Market/ GDP Ratio	Market Evolution Factor	Cumulative % Gross Domestic Product	Cumulative Environmental
1	United States	165.5	5951.0	2.78%	3.96	29%	40%
2	Japan	65.3	2468.0	2.65%	3.36	41%	56%
3	Germany	36.4	1398.0	2.60%	3.98	48%	65%
4	France	20.2	1080.0	1.87%	3.22	53%	70%
5	United Kingdom	17.6	920.6	1.91%	3.43	57%	74%
6	Italy	15.0	1012.0	1.48%	2.74	62%	78%
7	Canada	10.8	537.1	2.01%	3.56	65%	81%
8	Netherlands	6.7	259.8	2.58%	3.86	66%	82%
9	Spain	6.2	514.9	1.20%	2.69	68%	84%
10	Australia	5.3	293.5	1.81%	3.56	70%	85%
11	Sweden	4.3	145.6	2.95%	3.95	71%	86%
12	Switzerland	4.3	152.3	2.82%	3.74	71%	87%
13	Belgium	4.2	177.9	2.36%	3.67	72%	88%
14	Former USSR	4.0	479.0	0.84%	1.80	74%	89%
15	South Korea	3.4	287.0	1.18%	2.61	76%	90%
16	Austria	3.2	141.3	2.26%	3.32	77%	91%
17	Taiwan	3.1	209.0	1.48%	2.58	78%	92%
18	Denmark	2.8	94.2	2.97%	3.86	78%	92%
19	Brazil	2.5	369.0	0.68%	2.18	80%	93%
20	Norway	2.1	76.1	2.76%	3.67	80%	94%
21	Mexico	2.0	328.0	0.61%	2.31	82%	94%
22	China	1.6	506.1	0.32%	1.58	84%	94%
23	Finland	1.5	79.4	1.89%	3.22	85%	95%
24	Turkey	1.3	219.0	0.59%	2.14	86%	95%
25	Hong Kong	1.0	86.0	1.16%	3.14	86%	95%
26	Thailand	1.0	103.0	0.97%	1.97	87%	96%
27	Poland	1.0	167.6	0.60%	1.96	87%	96%
28	India	1.0	240.0	0.42%	1.53	89%	96%
29	New Zealand	0.9	49.8	1.81%	3.44	89%	96%
30	Greece	0.9	82.9	1.09%	2.72	89%	97%
31	Indonesia	0.9	133.0	0.68%	1.92	90%	97%
32	Singapore	0.8	45.9	1.74%	3.58	90%	97%
33	Portugal	0.7	93.7	0.75%	2.69	90%	97%
34	South Africa	0.7	115.0	0.61%	2.25	91%	97%
35	Argentina	0.7	112.0	0.63%	2.11	92%	97%
36	Malaysia	0.7	54.5	1.28%	1.97	92%	98%
37	Saudi Arabia	0.6	111.0	0.54%	2.19	92%	98%
38	Ireland	0.5	42.4	1.18%	2.69	93%	98%
39	Israel	0.5	57.4	0.87%	2.20	93%	98%
40	Iran	0.5	90.0	0.56%	1.56	93%	98%
41	Philippines	0.4	54.1	0.74%	1.81	94%	98%
42	Puerto Rico	0.4	22.8	1.54%	3.06	94%	98%
43	Chile	0.3	34.7	0.86%	2.27	94%	98%
44	Czech Republic	0.3	75.3	0.40%	2.20	94%	98%
45	Venezuela	0.3	57.8	0.52%	1.94	94%	99%
46	Hungary	0.3	55.4	0.54%	1.87	98%	99%
47	Colombia	0.3	51.0	0.59%	1.83	95%	99%
48	Romania	0.3	6304.0	0.47%	1.64	95%	99%
49	Egypt	0.3	41.2	0.68%	2.18	96%	99%
50	Peru	0.2	25.0	0.88%	1.83	96%	99%
	Rest of World	3.5	894.0	0.39%	1.73		
	Total World	**408.0**	**20658.0**	**1.98%**	**1.97**	**100%**	**100%**

Source: EBJ Journal 1995, Environmental Business International, Inc. Publishing Co., (revenues in $ billions).

Note: E.I. Market Evolution Factor is based on 17b weighted variables in market, regulatory, and social factors.

- Multilateral agencies and international financial institutions
- Foundations and nonprofit organizations
- Foreign governmental institutions
- Multinational corporations

International Development Assistance Agencies. The growing concern about the threat to the global environment has led international development agencies to focus attention on the environmental impacts of development projects they support. In addition to its policies on trade, foreign investment, and debt, the international development assistance community has increased its focus on global environmental concerns, especially in light of the activities proposed under the 1992 UNCED in Rio de Janierio and the Agenda 21 developed during the conference. Funding for sustainable development activities in developing countries is estimated by the UNCED secretariat to be more than $600 billion, including approximately $125 billion in grants or concessions from the international donor community. In addition to regular financial assistance appropriations by developed countries, it is very likely that the total available funding for assistance to developing countries will increase. For example, Figure 6.2 indicates that the total financial assistance

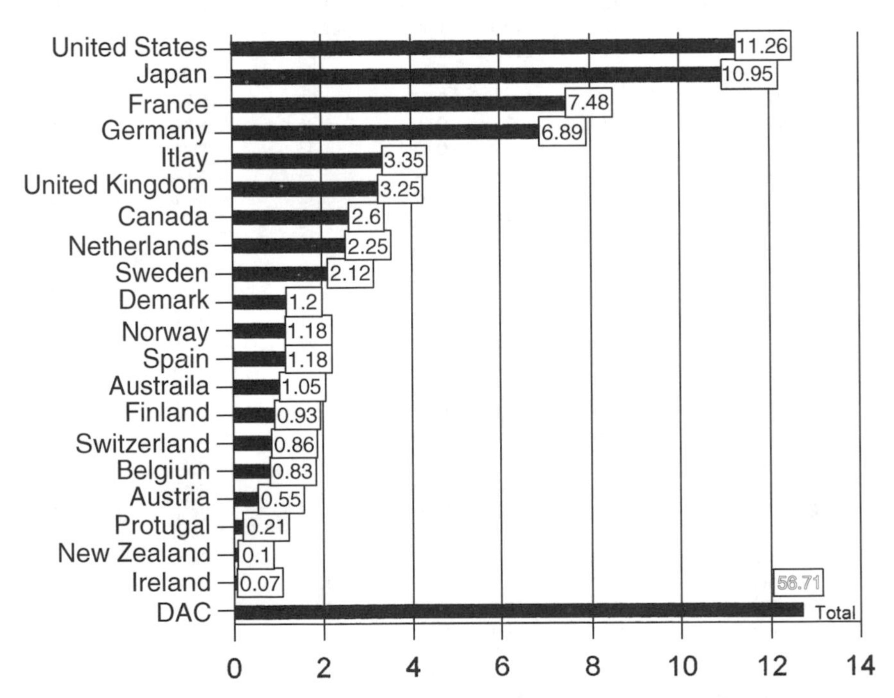

Figure 6.2 Net official development assistance from Development Assistance Countries (DACs), 1991 (*Source:* Organization for Economic Cooperation and Development (OECD), *Development Cooperation: 1992 Report* (OECD, Paris, 1992), Chart 2, p. 24)

for developing countries in 1991 alone ($57 billion) is approximately 10 times lower than the amount ($600 billion) estimated by the UNCED to implement the activities under Agenda 21. The World Bank has also indicated the need for approximately $75 billion annually to implement the high-priority items of global environmental and social programs. Many international consulting opportunities will be generated from these activities.

Multilateral Agencies and International Financial Institutions. The recent surge within the international community regarding the importance of sustainable development has brought about increased attention and redirection of the international multilateral and financial agencies to global environmental issues. Thus, there is a need for these institutions to provide financial and technical assistance for sound environmental management worldwide. The challenges are extremely tough in the light of the recent protests in many donor countries, which seek to shift foreign assistance to local domestic issues. Although these protests in major donor nations like the United States are likely to cause a decrease in financial assistance to many developing nations, there is optimism that global environmental issues will continue to loom large and will increase the demand for environmental consulting and practice in developing countries. Opportunities for environmental consulting continue to exist within the core activities of the international multilateral agencies and financial institutions.

Foundations and Nonprofit Organizations. The recent high rate of emergencies such as famine, drought, fires, and earthquakes and the deterioration of natural resources have increased the activities of nonprofit organizations and foundations in seeking environmental consultants for preliminary assessments, feasibility studies, and project design. Traditionally, foundations provide grants to a number of environmental projects, and in many cases they seek environmental consultants to review proposals and, sometimes, to oversee projects. To the extent that many nongovernmental organizations (NGOs) and foundations have weaknesses in technical resource capacity, they rely on consultants for technical assistance. The demand for technical support for NGOs and foundations will continue to grow. Opportunities are increasing for environmental consulting activities as NGOs and foundations continue to play a vital role in development assistance and the politics of development. In 1989 alone, private voluntary agencies spent nearly $4 billion, which represented about 5 percent of the total development assistance funds available globally. This amount is likely to quadruple as the demand for NGOs to participate in the design and supervision of donor institutions increases. This will generate a ripple effect, increasing the demand for international environmental consulting by NGOs.

Foreign Governmental Institutions. Government institutions are the entities that develop regulations and standards for the public. They are entrusted to implement these standards and regulations to protect human health and the environment. The government is a premier source for environmental contracts. In developing countries, governments often receive grants to implement environmental projects. These projects are mostly funded by the international donor community, international financial institutions, and international development agencies. The governments in turn send out Requests for Proposals (RFPs) and bids for the projects. Most of the foreign grants received by governments in developing countries have what is called *Tied Aid.* This means there are provisions by the grantor to have a portion of the contract work set aside for consultants from the grantor's country. This has always been and will continue to be a promising opportunity for international consulting practice in developing countries.

Multinational Corporations. In the last twenty years multinational corporations have increased their corporate environmental policies and procedures amid the growing pressures from the public as a result of industrial accidents such as the Bhopal, India disaster, the Exxon Valdez oil spill, Chernobyl. Multinational companies have become more environmental conscious and continue to focus their attention on green technologies and on pollution prevention.

As we move towards the year 2000, multinational corporations will increase their funding to environmental consultation in areas of manufacturing, production and management strategies, towards waste minimization and pollution prevention. Increase in environmental services such as waste management will continue to be a major sector that multinational companies will allocate funds and resources.

As many multinational companies enter new markets in developing countries such as Africa, Asia, and Eastern Europe, the pressure placed by the right-wing environmental organization and in some cases in-country environmental organizations will provide a new breed of environmental management business from the multinational companies.

Clean-up of past activities in many developing countries is on the rise. These clean-ups are being fueled by the massive negative impact to human and the natural resources created as a result of releases of hazardous waste and hazardous constituents to the environment by multinational corporations. Henceforth, environmental remediation services will be in high demand in many regions within the developing countries.

Multinational companies will spend half the total amount of environmental spending by the year 2000 on nongovernmental regulatory activities. Most of the environmental expenditures will be in anti-pollution and green-products research and development activities.

6.1.1 Africa

The primary causes of environmental degradation in Africa are shown in Figures 6.3. It provides a summary of major environmental concerns in African countries. The combination of social and economic decline has exerted tremendous pressure on natural resources and the environment, leading to massive degradation in all sectors of the environment. Urbanization and pov-

Figure 6.3 Primary causes of environmental deterioration in regions of Africa. *Developed by:* African Environmental Research and Consulting Group (a nonprofit organization), 14912 Walmer Street, Overland Park, KS 66223, USA. Tel/Fax 913/897-6132. Survey/Research: AERCG, February 1994, Coordinator: Peter A. Sam, Asc, BS, MS.

erty are widespread on the continent and, in many cases, have led to overcrowding, unsanitary conditions, and minimal basic public amenities. The lack of adequate environmental policies, regulations, and standards has also contributed to the indiscriminate disposal of hazardous waste and toxic chemicals by industries. Lack of environmental policing and enforcement are also contributing factors to the deteriorating environmental conditions on the continent. The sectors that constitute the lion's share of the present environmental market in Africa are water quality and supply, sewage treatment, municipal solid waste management, waste management equipment, pollution control technology and equipment, and institutional environmental capacity building (training, development of expertise, private and public institutional development). The African Development Bank estimates that 73 percent of the African environmental market at present is within the water quality and supply, sewage treatment and disposal sectors, and approximately 36 percent of the current market is attributed to pollution abatement control technology and equipment.

Africa faces immense environmental and natural resource management challenges. Rapid population growth, poverty, and unsanitary living conditions are causing widespread and growing environmental damage. International financing and development aid institutions have incorporated environmental issues into project designs, which provides opportunities for environmental consultants. The nexus between economic development and environmental management has brought about increased demand for environmental consulting whenever the development aid is being considered for virtually any economic development project. For instance, the World Bank's National Environmental Action Plans (NEAPs) in their economic development and environmental programs (Structural Adjustment Programs) require developing countries, especially Africa, to develop their own NEAPs. The NEAP helps to define key environmental problems and issues within a country and to prioritize these issues for remediation. The NEAP is a key information tool for major environmental priorities in a country. Since 1990, the African Development Bank has instituted environmental screening and categorization in accordance with the bank's environmental guidelines for every project it finances. These requirements ensure that projects are environmentally sound and do not have negative impacts on the environment and human health. Most consulting opportunities will generate from EIAs and EISs as economic revitalization efforts expand. There will also be many opportunities for environmental consulting as a result of the World Bank's NEAP in many African countries. There are other programs and activities that provide potential markets for environmental consulting, such as the Global Consensus on Sustainable Development, the Environmental Impact Assessment (EIA) requirement by international donor/developing aid agencies and international financial institutions for any capital improvement projects, the Global Environmental Facility (GEF) program, and the recent United Nations' $30 billion Special Initiative for Africa.

6.1.2 Eastern Europe

As a result of decades of unsound environmental management under the Communist rule, Eastern Europe has witnessed a decline in human health such as evidenced by high infant mortality rates and high levels of lead in children. Inadequate environmental policy and standards have led to poor air quality and the indiscriminate dumping of hazardous and toxic chemicals. Lack of appropriate management and low standards of industrial waste disposal have been due to lack of expertise, lack of regulators, lack of technology, and in some countries within the region, lack of funding to provide sound environmental management. The transition to economic and political reforms has, to an extent, been successful and has directed some attention to environmental policies, especially in setting standards for industrial pollution control and waste management. This approach to environmental stewardship through environmental legislation and rules development is slow at best. The move toward a European Union is providing a catalyst for countries in the region such as Poland, Hungary, and the Czech Republic to make quick reforms in their economic and environmental agendas and to move closer and/or to meet the European Union's overall stated objectives. International developmental assistance institutions and agencies have also served as catalysts in addressing environmental problems and seeking adequate management, through their insistence on environmental impact assessments as a condition in providing assistance in economic restructuring and development. Examples are the Organization for Economic Cooperation and Development (OECD) community, which provides development assistance and aid to the countries in the region, and the multilateral banks such as the European Bank for Reconstruction and Development. The region lacks enforcement of environmental regulations in many of its countries. This has been a leading contributing factor to the massive environmental damage that has occurred over the years in Eastern Europe. Civil war and ethnic conflict in the region—for example, in Poland—have caused severe environmental damage and continued to cause a massive loss of natural resources, including loss of biodiversity.

Environmental consulting prospects in Eastern Europe are great, and there is tremendous pressure on governments to prioritize environmental remediation and management. In 1995 four of the countries in the region (Poland, the Czech Republic, Romania and Hungary) were listed in the top 50 environmental markets in the world by the *Environmental Business Journal* vol. VIII; #8, 1995.

There is much assistance to this region by donor countries such as the United States. Several projects are being initiated by the governments in the region, which seek consultation from the West, especially from the U.S. environmental experts and companies, to assist in environmental management, design, construction, and environmental institutional capacity building. The future for environmental consulting in the region is promising.

6.1.3 Western Europe

Western European countries may face increasing environmental problems if regional environmental regulations, standards, and policies are not instituted along with the proposed European Community coalition on trade and economy. Without appropriate management, the free flow of goods will lead to the free flow of pollutants. The regional environmental problems are mainly coastal and surface water pollution. Sources of pollutants are poorly managed municipal waste and inadequate sewage management systems, resulting in surface water and coastal water pollution. For example, the Meuse River collects refuse and carries it through France, Belgium, and the Netherlands to its final entry into the North Sea. It carries solid and industrial waste that may have been disposed of into surface waters. The emergence of the European Community (EC) coalition will force member-nations to develop strict environmental policy and standards. This will generate markets for environmental consultants as the EC formalizes and finalizes plans for the future union of its trade and economic coalition. The growth range for this promising environmental market is estimated at an average of 3 to 6 percent for each country in the region. Germany is leading the region in market size at $36.4 billion, with France at $20.2 billion, the United Kingdom $17.6 billion, Italy $15.0 billion, Netherlands $6.7 billion, Spain $6.1 billion, Sweden $4.8 billion, Switzerland $4.7 billion, Belgium $4.2 billion, Austria $3.2 billion, and the rest of the countries in the region register a total of $8.5 billion in the market. The region will maintain a growing environmental industry, especially in pollution prevention/waste minimization, wastewater treatment, air pollution abatement, and environmental remediation and restoration.

6.1.4 Latin America

There are various factors that indicate potential growth for the environmental market in Latin America. For example, the passage of the North American Free Trade Agreement (NAFTA) brought about drastic changes in Mexico's approach to environmental policy and management. In Central America, the peace treaty to end the war between Nicaragua and El Salvador has led to a new start by governments in the region to focus on biodiversity and reclaim the vast rain forest. Foreign investors and companies in the pharmaceutical industry are beginning to invest in biodiversity and preservation of potential plant species for pharmacological purposes. The environmental market for the entire Latin American region reached approximately $6.7 billion in 1994 (according to *Environmental Business International Journal,* San Diego, California), and was projected to grow at 10 percent annually to exceed $10 billion by the end of 1998 and projected to reach $13 billion in 2000. The most promising segment of environmental requirement within the region is the need for pollution control technologies and equipment. Of next importance is the need for comprehensive municipal environmental planning, design, and construction of solid waste management systems. Treatment of waste, both domestic and industrial, is virtually inadequate in most countries in the region. Great opportunities lie in the design and construction of sewage treatment systems and in provision of clean drinking water. The areas of importance to human health and environmental protection lie in infrastructure redesign and construction related to landuse planning and design, waste management systems, water treatment plants and water delivery, domestic sewage treatment, and management of industrial waste. Air pollution is a major problem in the region and, like biodiversity and forest preservation, is an important segment of the environmental market. Control of emissions from all sources, especially industries and automobiles, is of crucial concern owing to rising urbanization and population growth in urban centers.

Brazil leads the region with $2.4 billion of the environmental market, followed by Mexico with approximately $2.0 billion. Mexico experienced a slowdown in its environmental market when it underwent economic crisis, but with NAFTA in place, the United States, in cooperation with the Mexican government, has made funds available that will drive the market up again, especially in areas such as water infrastructure, wastewater treatment equipment, and hazardous waste management and remediation.

Other countries in the region are showing positive signs for an increase in environmental markets, as they restructure their economies through implementation of Structural Adjustment Programs and privatization programs/projects. As institution of environmental policies and legislation continues in most countries of Latin America, environmental markets will continue to emerge and their growth will outpace other developments in the region. In Chile, the passage of the Chilean Environmental Framework Law in March of 1994 has begun to yield signs of a demand for environmental goods and services. The Brazilian government initiative in economic growth through encouraging foreign investment has shown signs of growth within the wastewater treatment sector, which is now enjoying increased spending.

One shining area of the environmental market in the region is the need for both public and private sector environmental capacity building. The need for experts in environmental management and enforcement of environmental laws and standards is the impetus for the environmental market to grow at a faster pace. International environmental consulting firms and individual consultants with expertise in environmental training and education have great opportunities at the present time. Future opportunities are wide open, but priority areas of growth and demand are municipal solid waste management, sewage treatment, design and construction, water treatment and water supply, pollution abatement, and control technologies.

6.1.5 Asia

Much of Asia has exhibited tremendous economic growth, which has outpaced the control and management of pollution. As a result, rapid industrialization and population growth have led to the pollution of surface water (lakes, rivers, and coastal waters), air, and soil. Countries in east Asia exemplify this situation. The southern region of Asia, however, has had little to modest economic growth, which has consequently led to a high rate of poverty and high population density in the urban centers, resulting in massive pressure on its natural resources. Pakistan, Sri Lanka, India, and Afghanistan are examples of countries in the region exhibiting ecological deterioration owing, in part, to poverty, population growth, and economic stagnation. Deforestation is prominent in the southeast region of the continent, where most of the land is cultivated without reforestation. In some countries in the region, for example, in the Philippines, exploitation of the forest for lumber has taken its toll on the natural resources.

The continent of Asia is growing fast, and there are promising environmental consulting opportunities in industrial pollution abatement and control, agro-forestry, municipal solid waste management, environmental policy, and capacity building. The poorer countries in the region will continue to strengthen their environmental laws, regulations, and management; therefore, the opportunities within these countries will be more promising than those in richer countries in the region, such as Japan. Japan's environmental market ranged from $64.2 billion in 1993 to approximately $66 billion in 1994 and projected to reach $81.7 billion in 2000. The rest of Asia will capture a net market of $26.6 billion in 1998 and project to reach $36.4 billion in 2000. There are opportunities in Japan, but foreign business penetration into the Japanese market is difficult because of protectionism. Environmental problems such as municipal waste management, adequate human shelter, sanitation and waste treatment, that face this country will open doors to the Japanese environmental market for environmental experts and firms with innovative solutions and approaches to urban infrastructure development, municipal waste management, and environmental restoration and remediation.

6.1.6 Pacific Rim

The main environmental problem of the Pacific Rim is the threat to the coastal natural resources and ecology. Management of municipal solid waste and sewage disposal are critical to comprehensive coastal management. There are promising consulting opportunities, mainly in coastal and solid waste management. Most governments in the region are beginning to strengthen their environmental policies and regulations. As these governments seek assistance in building their environmental management capacity within their various institutions, there will be a growing market for consulting as the regional environmental market grows at approximately 5% per annum. In 1998, the market reached $7.5 billion and projected to reach $8.3 billion by 2000.

6.2 LENDING FOR ENVIRONMENTAL MANAGEMENT

Most of the countries where there are good prospects for environmental consulting are poor; therefore, the governments depend on foreign aid and assistance in environmental management and protection. Financing is of key importance for countries that seek to undertake sound environmental management and environmental stewardship. The level of financing dictates the need for environmental consulting, because the funds will be available to hire consultants and guarantee payments for services. The international multilateral banks and development aid agencies have steadily increased their fi-

Table 6.2 Lending Portfolio for Environmental Management and Pollution Control Approved in Fiscal 1989–93 (millions of dollars)

Country	Project	Loan/Credit Amount
	New Commitments, Approved in Fiscal 1993	
Brazil	Water Quality and Pollution Control, Sao Paulo/Parana	245.0
	Minas Gerais Water Quality and Pollution Control	
China	Southern Jiagsu Environmental Protection	145.0
India	Renewable Resources Development	250.0
Korea, Republic of	Kwangju and Seoul Sewerage	190.0
Mexico	Transport Air Quality Management	110.0
Turkey	Bursa Water and Sanitation	220.0
Total		1,289.5
	Projects Under Implementation, Approved in Fiscal 1989–92	
Angola	Lobito-Benguela Urban Environmental Rehabilitation (92)	46.0
Brazil	National Industrial Pollution Control (92)	50.0
Chile	Second Valparaiso Water Supply and Sewerage (91)	50.0
China	Ship Waste Disposal (92)	15.0
	Beijing Environmental (92)	125.0
	Tiajin Urban Development and Environmental (92)	100.0
Côte d'Ivoire	Abidjan Lagoon Environment Protection (90)	21.9
Czech and Slovak Federal Republics	Power and Environmental Improvement (92)	246.0
India	Industrial Pollution Control (91)	155.6
Korea, Republic of	Pusan and Taejon Sewerage (92)	40.0
Poland	Energy Resources Development (90)	250.0
	Heat Supply Restructuring and Conservation (91)	340.0
Total		1,439.5
Portfolio Total		2,729.0

Source: World Development Report 1993, World Bank Publication.

Table 6.3 The GEF Biodiversity Grants

Africa	
UNDP	
Burkina Faso	2.5
Côte d'Ivoire	3.0
East Africa (institutions)	10.0
Ethiopia	2.5
Mauritius	0.2
Lake Tanganyika	10.0
West and Central Africa	1.0
WORLD BANK	
Cameron	5.0
Central Africa	1.8
Congo	10.0
Ghana	7.2
Kenya	6.2
Malawi	4.0
Mozambique	1.8
Uganda	4.0
West Africa (game ranching)	7.0
Zimbabwe	5.0
Asia and the Pacific	
UNDP	
Indonesia and Malaysia	2.0
Mongolia	1.5
Nepal	3.8
Papua New Guinea	5.0
South Pacific	8.2
Sri Lanka	4.1
Vietnam	3.0
WORLD BANK	
Bhutan	10.0
Indonesia	12.0
Laos	5.5
Philippines	20.0
Subtotal	75.1
Jordan	6.3
Yemen	2.8
Latin America and the Caribbean	
UNDP	
Amazon Region	4.5
Argentina	2.8
Belize	3.0
Colombia	9.0
Costa Rica	8.0
Cube	
Dominican Republic	3.0
Guyana	3.0
Uruguay	3.0
WORLD BANK	
Boliva	4.5
Brazil	30.0
Ecuador	6.0
Mexico	30.0
Peru	4.0
Subtotal	112.8
Eastern Europe	
WORLD BANK	
Poland	4.5
Romania and Ukraine	6.0
Turkey	5.0
UNEP	
Country Studies I	4.8
Country Studies II	2.5
Global Assessment	2.0
Data Management	3.5
Subtotal	12.8
Total	327.3

nancing of environmental projects. For example, in 1993 alone, the World Bank approved loans and credits for seven projects involving roughly $1.3 billion for pollution control, ten projects totaling $521 million for natural resources management, and six projects involving approximately $173 million for environmental institution capacity building. Nine projects in Asia were approved by the World Bank to the tune of $830 million; $681 million for six projects in Latin America and the Caribbean; $428 million for five projects in the Middle East, and North Africa, Europe, Central Asia; and $45 million for three projects in sub-Saharan Africa. This funding level will remain steady well into the next century; therefore, the opportunities and prospects for environmental consulting will continue to grow. Table 6.2 shows the lending record of the World Bank between 1989 and 1993. All indications are that this level of funding by the World Bank will continue in the foreseeable future.

It is fair to assume that steady lending for financing environmental projects will provide a steady demand for international environmental consultants. The opportunities will persist. The next section of this chapter examines an example of global financing programs that serve as the major drives for initiating environmental projects (the Global Environmental Facility). It provides information on future opportunities and prospects for international environmental consultants.

6.3 GLOBAL ENVIRONMENTAL FACILITY

The Global Environmental Facility (GEF) provides a promising prospect for future environmental consulting opportunities because it offers funding for environmental projects, which gives in-country governments the means to hire and guarantee payment of services to international environmental consultants. The GEF was established in 1990 as a three-year pilot project administered jointly by the World Bank, United Nations Development Programme, and the United Nations Environmental Programme. The first phase of the project ended in 1993, but disbursement of funds was anticipated to continue well into 2000. Initially committed GEF funds amounted to $1.3 billion. The Global Environmental Trust Fund, which is the core fund of the GEF, and associated cofinancing funds committed an additional $1.1 billion to GEF by 1993. Overall, less than 10 percent of these monies were spent during this period. Many countries have received grants under GEF for environmental undertakings such as biodiversity projects. Table 6.3 lists the GEF Biodiversity Grants awarded or received by various countries. The GEF program will continue to provide excellent opportunities and prospects for future environmental consulting, especially in Eastern Europe, Africa, and Asia.

7

AS WE APPROACH THE YEAR 2000 AND BEYOND

Global environmental problems will increase if the global community at large does not take steps to manage the world's natural resources. Among the critical environmental issues is the present rate of urbanization, which can have a severe impact on the biological, physical, geological, and chemical components of the earth and can be very costly. The future challenges were discussed at the 1992 United Nations Conference on Environment and Development (UNCED) held in Rio de Janeiro, Brazil. At the conference world leaders agreed to be responsible for global environmental management, natural resource management, and sustainable development. This agreement became the centerpiece of the Global Agenda 21, which paved the way for global environmental management, including conservation and sustainable development, for the year 2000 and beyond.

7.1 PROMISING FUTURE

The Global Agenda 21, an agenda for the next century, provides a blueprint for future global action in high-priority environmental and development areas and for legal instruments addressing such issues as climate change, biodiversity, and desertification. It establishes a working framework and includes provisions such as the following:

- Financing to protect natural habitat and biodiversity
- Allocating funds for international aid programs with high returns in the areas of poverty alleviation and environmental health (providing sanita-

tion and clean water, reducing indoor air pollution, and meeting basic needs)
- Investing in research and extension programs to reduce soil erosion and degradation and to put agricultural practices on a sustainable path
- Increasing resources for family planning and for primary and secondary education, particularly for girls and women
- Supporting governments in their attempts to remove macroeconomic imbalances that damage the environment
- Investing in research and development of non-carbon-based energy alternatives to respond to global climate change
- Eradicating protectionist pressures and ensuring that international markets for goods and services, including finance and technology, remain open.

Another promising element within the global environmental arena is the evolution of environmental standards. The International Organization for Standardization (ISO) has developed a series of international standards, which are growing in popularity and becoming universal standards among businesses and governments worldwide. For example, ISO 14000 provides basic standards for environmental management systems, environmental auditing, environmental labeling, environmental performance management, and product life cycle assessment. The growing global environmental awareness within grass roots organizations, within governments, and among citizens has led to similar awareness within corporate business throughout the world, especially within multinational corporations. Corporate CEOs and stockholders have become more conscious and increasingly proactive in regard to corporate environmental management. Figure 7.1 illustrates the direction of multinational corporations for the future, which indicates promising markets for international environmental consulting.

7.2 EMERGING ENVIRONMENTAL FIELDS

As much as one half of the global environmental market is likely to shift to nonregulatory segments as opposed to those subject to environmental laws and regulations. The following are areas that show promise for growth as we move into the twenty-first century, expected to account for at least half of the global environmental market by the year 2000:

- Strategic environmental management
- Environmental responsibility programs—International Standards Organization (ISO 14000)—and Pollution Prevention
- Environmental technologies: pollution prevention control equipment

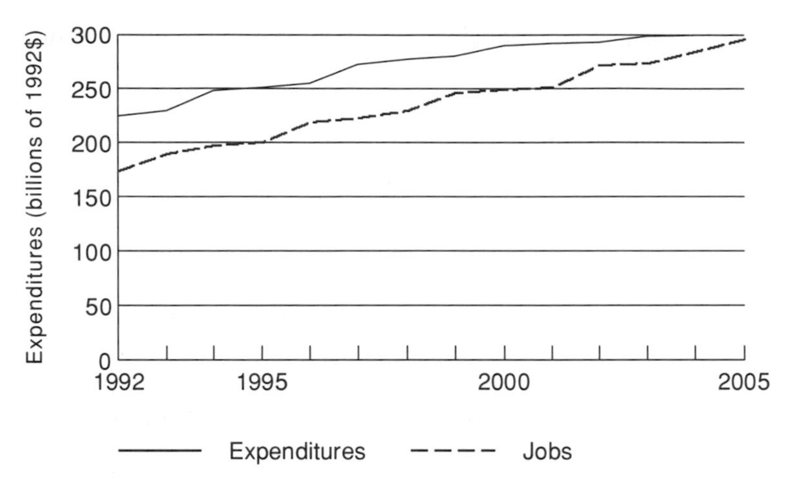

Figure 7.1 Growth of environmental protection spending and jobs, 1992–2005 (*Source:* Forecast by Management Information Service, Inc., March 1993)

- Environmental medicine/plant pharmaceuticals
- Ecotourism
- Occupational health

7.2.1 Strategic Environmental Management

Whereas market opportunities previously lay in compliance with environmental regulations, the concepts of sustainable development and pollution prevention have emerged as a holistic approach for developing countries to address environmental and socioeconomic problems. These are the new concepts of environmental management that are evolving globally. Corporations proactive approach to management of their waste toward a competitive advantage is a growing concept in industrialized countries such as the United States. The regulated universe is beginning to realize the need to be proactive and innovative in order to stay competitive as its members face mountains of federal, state, and local environmental regulations and policies. Corporations find the new approach quite realistic because of its apparent benefits and, especially, because of the uncertainties surrounding future environmental regulations and standards. This new approach to environmental management is called *Strategic Environmental Management* (SEM), which can be defined as the process of linking environmental performance to corporate overall economic performance toward a sustainable business practice.

In the past corporations have responded and reacted to environmental regulations and operated in the compliance mode. The shift to SEM is a cost-effective/cost-saving approach to integrating waste management with the overall mission of the corporation. The integration of waste management within a corporation can be viewed as an investment, as has been the case in many companies in the United States that have developed environmental policies and shifted to investing in pollution prevention (which entails source reduction, resource recovery, recycling, and product substitution), product redesign, "environment-friendly" packaging, and environmentally good housekeeping. SEM is an emerging environmental field and the wave of the future in several countries. A number of initiatives in the United States toward this novel concept, which has been described as a commonsense approach, are currently under way. Among these are the following:

- Community Partnership Initiative
- EPA's Voluntary Compliance Initiatives
- Voluntary RCRA Corrective Action

SEM and sustainable development principles and theories will be the future tools for effective corporate environmental management. SEM promotes a corporation's image as a good environmental steward and provides a competitive approach to doing business. SEM and sustainable development will grow steadily as we approach the year 2000.

7.2.2 Environmental Responsibility Programs—International Standards Organization (ISO 14000) and Pollution Prevention

The concept of Environmental Responsibility Programs (ERP) evolved out of the notion of solving environmental problems before they occur. The "end-of-pipe" traditional approach is becoming obsolete and the commonsense approach, as it is called, is fostering a new market in which corporations around the world are seeking "green business." Figure 7.1 illustrates the trend of multinational companies in generating environmental markets through innovation. Green business generally means doing business in an environmentally responsible and cost-effective manner. ERP is an emerging market, already gaining momentum in the United States and Western Europe.

ERP is a nonregulatory, proactive, and innovative approach whereby companies seek to restructure and redesign their manufacturing processes, alter products to reduce the amount of pollutants they generate, and operate in an environmentally safe manner. The impetus for ERP was the realization that it costs more to clean up than to prevent pollution and, thus, the desire to avoid the high cost of environmental capital expenditures. The other factor underlying ERP promotion is the recent consumer pressure on companies to be environmentally responsible in their product packaging and activities, which has had a tremendous impact on corporations' environmental purchasing and marketing decisions, especially in the developed countries. It is becoming evident that consumers are more reluctant to purchase "non-green" products; therefore, manufactures and corporations around the world are responding and seeking ERP approaches. Corporations now know that to be competitive and remain financially successful in the present and future global market, it is imperative to be environmentally responsible and to be portrayed as good environmental stewards. ERP is indeed emerging as the commonsense approach to doing business as we near the year 2000. It provides corporations, and other entities, with benefits and advantages, such as the following:

- Cost-effective ways of doing business by saving money through the up-front reduction of waste treatment, disposal costs, raw material purchases, and other environmental compliance costs
- The elimination or reduction of environmental liabilities
- Safer working environments, protecting public health and workers' health and safety
- Protection of the environment

The two major drivers for ERP are ISO 14000 and Pollution Prevention (P2).

ISO 14000. As described in Chapter 3, in the section entitled "Complex Forms of Registration and Licensing", ISO 14000 is an important element of ERP because it is intended to establish a universal standard for environmental management. The ISO itself provides a series of Environmental Management Systems (EMS) which offer specific approaches to holistic environmental management. ISO 14000 comprises:

- Life cycle analysis
- Environmental labeling and marking
- Environmental assessments and auditing
- Environmental management systems

Pollution Prevention. The Pollution Prevention (P2) movement has come to the forefront in the global environmental policy or agenda of the 1990s. P2 entails the reduction or elimination of pollutants at the source so that waste is not generated, in contrasts to "end-of-pipe," "collect-and contain," and "release-and-dilute" controls designed to treat or control releases and wastes already generated. P2, as defined by the U.S. Environmental Protection Agency under the Pollution Prevention Act of 1990, is the maximum feasible reduction of all wastes generated at production sites. It involves the judicious use of resources through source reduction, energy efficiency, reuse of input materials during production, and reduced water consumption. Figures 7.2a and 7.2b illustrate the environmental management options hierarchy for pollution prevention and source reduction methods that are considered the optimal approach to P2.

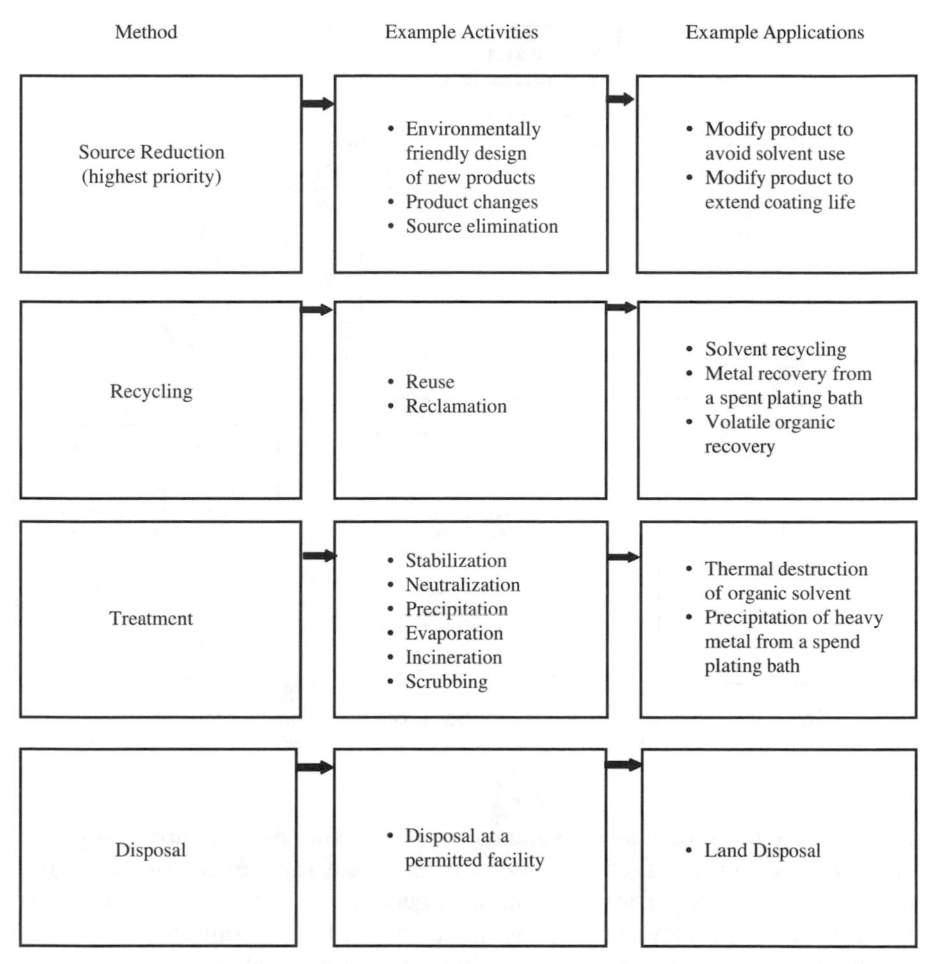

Figure 7.2a Environmental management options hierarchy (*Source: U.S. Environmental Protection Agency, Facility Pollution Prevention Guide,* May 1992)

7.2.3 Environmental Medicine

Environmental medicine, as a market segment, has become an increasingly important driving force in forest preservation and conservation and in biodiversity. Medicinal plants have been sources of remedies for many diseases in communities worldwide where traditional medicine is used by native peoples. Currently, an increasing number of people are turning to traditional medicine. Most pharmaceutical companies are heavily investing in botanical medicines and, as a result, are recruiting environmental practitioners to assist in promoting sustainable harvesting of plant species and the natural forest ecology at large. Environmental medicine has increased the demand for in-

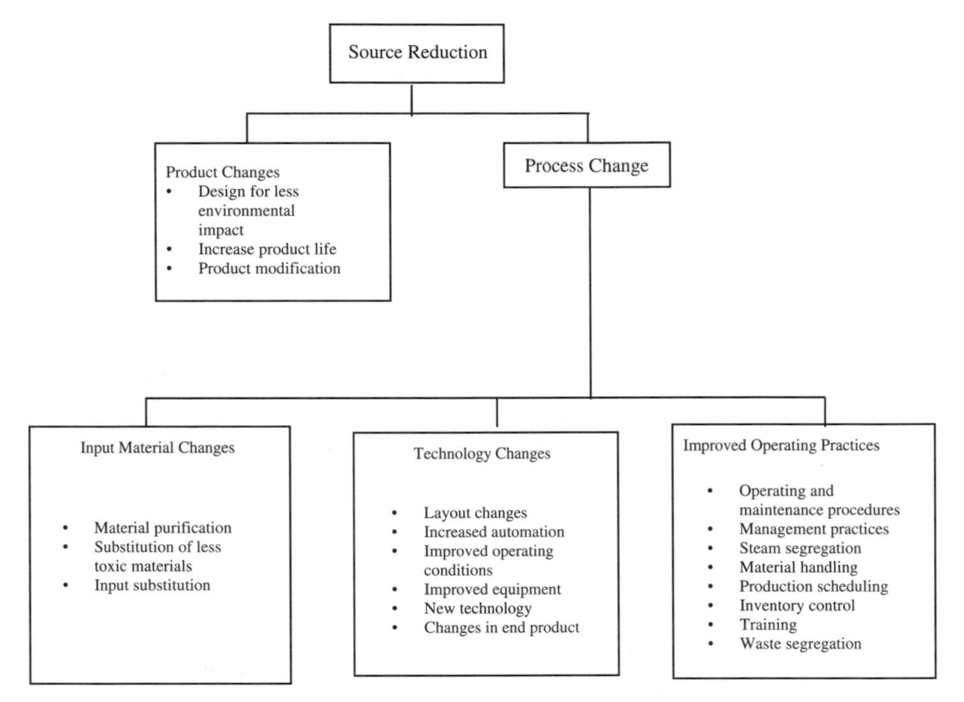

Figure 7.2b Source reduction methods (*Source: U.S. Environmental Protection Agency, Facility Pollution Prevention Guide,* May 1992)

ternational environmental consultants on forest and plant pharmacology in developing countries, such as those with the great rain forests of the world, ranging from West Africa to Central America. Environmental medicine is relatively at its infancy and largely unknown, but it is becoming increasingly popular worldwide as pharmaceutical industries and investors seek to discover new plant species with potential medicinal values. Environmental consultants are engaged mainly to conduct feasibility studies and preliminary assessments of the impact of research and development activities on the rain forest regions in order to maintain biodiversity and sustainable management of plant species.

7.2.4 Ecotourism

The field of ecotourism, which emerged in the mid-1990s, is growing and offers promising opportunities for international environmental consultants in the West Indies, South Africa, Zimbabwe, and Western Europe. Ecotourism is increasing in popularity within the context of economic growth and development and natural resource sustainability. Developing countries will continue to experience increases in the ecotourism market; exposing visitors to their cultural heritage becoming a gateway to boosting the local economy. At the same time, within the concept of ecotourism, approaches and solutions can be developed to help indigenous people to disengage from subsistence practices that degrade the environment and lead to loss of biodiversity and cultural heritage. The growing market alternative, economic growth and preservation of unique biodiversity and wildlife inheritance, is contributing to the growth of ecotourism in developing countries. This segment will continue to seek and demand experts in environmental economics, environmental impact assessment, and environmental design (land use design). Driving forces for these areas of demand will occur within financial institutions, mainly for capital investment in hotels, and in international assistance/donor agencies funding socioeconomic and development projects. These agencies and institutions need third-party consultants to provide environmental impact assessments. As revenues from tourism grow in developing countries, it is anticipated that the market for ecotourism will also grow and the demand for ecotourism consulting will increase exponentially. Between 1989 and 1990 revenues from tourism jumped to 11.6 percent in South America, 8 percent in the Caribbean region, 11 percent in the East Asia/Pacific region, and 19 percent in the southern Africa region. Good examples of successfully implemented ecotourism projects are the Zimbabwe Campfire project and the Phinda Resources Reserve in Natal Province, South Africa.

7.2.5 Occupational Health

Occupational health is an area with potential to grow as we near the year 2000. The main reasons are increasing concern about workplace safety and the rising consciousness of employees about the health effects of the workplace. Workers' long hours spent using a computer and the structure and settings of a facility are investigative areas that call for international environmental consultants to assess the impact of workplace policies, settings, structure, and practices on the universe of the human work force. There are other areas more prominent in occupational health and safety in the industrial operational sector; here international environmental consultants are increasingly in demand to conduct environmental safety assessments to evaluate the effect of exposure to operations at facilities that generate hazardous waste and utilize toxic chemicals.

7.3 INTERNATIONAL ENVIRONMENTAL DATA AND TRENDS

The global environment and natural resources face unprecedented risks as we approach the year 2000 and beyond. The rate of urbanization, in addition to population explosion and poverty in many cities around the world, presents dynamic challenges to the sustainability of our global natural resources. World population is projected to grow by 3.7 billion between 1990 and 2030. The increase in global population and the rate of urbanization, which is projected to grow at 170,000 people per day, pose a threat to natural resources such as

Table 7.1 Countries Making Up the Highest Pollution Sources for Greenhouse Emissions in 1989

Country	Percent Emission
United States	18.4%
USSR	13.5%
China	8.4%
Japan	5.6%
Brazil	3.8%
India	3.5%

Source: An Environment SOS, *The China Business Review,* Sept–Oct 1992 Washington D.C. p. 31.

wetlands and other fragile ecological systems (with biophysical and chemical components) that are essential to our survival on earth. The threat to public health and the environment challenges the nations to address these negative environmental trends.

Data from a number of sources indicate threats to the global environment if appropriate actions by internationals communities are not pursued. The data indicate that industrialized nations account for the lion's share of greenhouse gasses and chlorofluorocarbons (CFCs). Tables 7.2 and 7.3 provide data on the countries registering the highest industrial emissions of carbon dioxide.

Greenhouse emissions in high income countries have registered the highest percentage of air pollution in the 1980s through 1990s. In 1989, as shown in Table 7.1, the U.S., U.S.S.R. recorded the highest air pollution emissions. Figure 7.3 indicates that low income countries are steadily increasing their air pollution emissions as they forge ahead to diversifying and privatizing the economies. Although the problem of air pollution is being addressed to a large extent in developed nations through environmental regulations, standards, and enforcement, this is not the case in less developed nations. There are several global and national actions being initiated by individual nations and the United Nations, including policy development, development of environmental legislation and standards, and global financing for specific en-

Table 7.2 Countries Registering the Highest Industrial Emissions of Carbon Dioxide in 1989

Country	Amounts in Tons
United States	4,869,005
USSR	3,804,001
China	2,388,613
Japan	1,040,554
Germany	964,028
India	651,936

Source: The China Business Review, Sept–Oct 1992 Edition 1992. Washington D.C.

Table 7.3 Countries with the Highest CFC Emissions in 1989

Country	Amounts in Tons
United States	130,000
Japan	95,000
USSR	67,000
Germany	34,000
Italy	25,000
United Kingdom	25,000
France	24,000
Spain	17,000
China	12,000

Source: World Bank, *World Development Report 1992.* A Compendium of Data on Global Change (The CO_2 Information Analysis Center, Oak Ridge National Laboratory, August 1990).

vironmental problems within regions worldwide, especially for the poor (low-income) nations. Substantial efforts by industrialized nations to provide funds and assistance to less developed countries to address the problem of air pollution offer promising opportunities for international environmental practitioners. The following are priority areas and issues of international concerns:

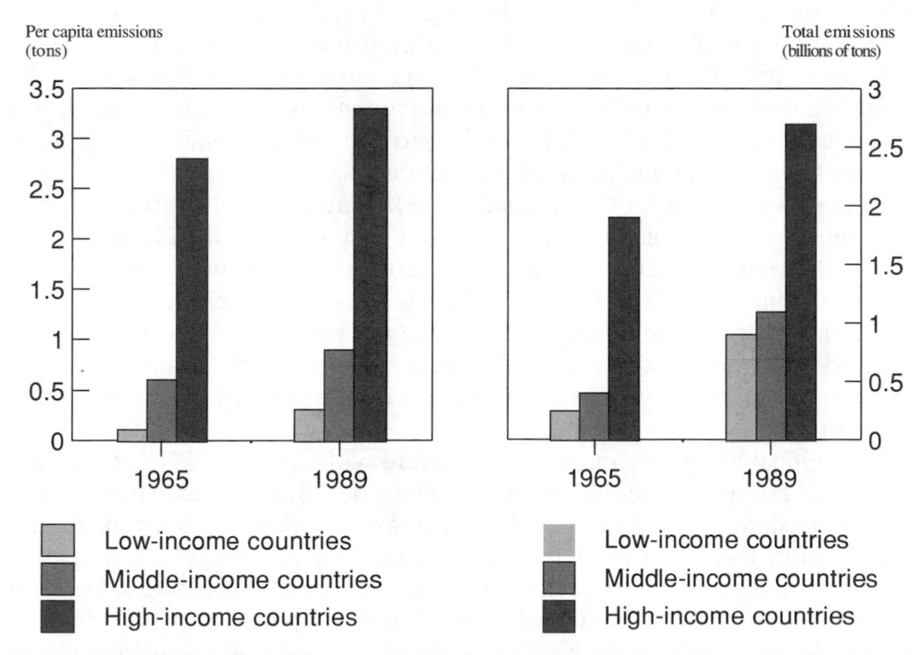

Figure 7.3 Global carbon dioxide emissions from fossil fuel consumption and cement manufacturing, 1965 and 1989 (*Source:* World Bank, *World Development Report,* Environmental Data, Appendix Table A.9, 1992)

- At least one third of the world's population has minimal or inadequate/unsafe drinking water and sanitation.
- Air pollution (such as caused by soot and smoke) presents unhealthy and unsafe health conditions to 1.3 billion people.
- Between 300 million and 700 million women and children suffer from various health problems owing to indoor air pollution—specifically, pollution caused by the use of firewood.
- Massive soil infertility and deforestation have resulted from unsustainable agricultural practices and cutting of trees.
- Increasing amounts of municipal solid waste have been generated in cities worldwide.
- Increasing urbanization in developing countries is associated with worsening urban poverty, infrastructure decay, and environmental degradation.
- Environmental standards and regulations in developing countries are inadequate.

7.3.1 Global Ambient Air Quality Indicators and Trends

Air pollution is the most significant and pressing global environmental issue at this time. Global warming and climate change are, in part, the results of air pollution, especially from greenhouse gas emissions. Major sources of global air pollution are found in heavily industrialized countries. However, increasing economic activities within the industrial sector in most countries have not employed appropriate air pollution abatement technologies. Even the industrialized nations with more stringent regulations still register high levels of air pollution. Tables 7.1, 7.2, and 7.3 provide statistics and data trends on air emissions for the highest producer nations.

The driving forces for the air quality market are the regulations established in industrialized nations such as the United States, where the Clean Air Act of 1990 provisions require stringent regulatory standards for various sources. In developing countries where air pollution laws and policies are either less stringent or absent, the driving force for the air quality market is the pressure exerted by international treaties and agreements, such as the Framework Convention on Climate Change, that have been signed by more than 160 countries.

The global air quality market is beckoning with opportunities and growing at a rapid rate in middle- to low-income countries. Some industrialized countries in Eastern Europe and Asia show increasing opportunities for the future. Increases in this segment of the global environmental market are propelled by commitments made by parties to the global treaties to reduce CFCs and greenhouse gas emissions. Air pollution control equipment is in great demand in developing countries such as those in Africa and Latin America and in countries within the Eastern Europe and Asia regions. In the air quality market in developing countries, demand is increasing in the areas of consulting, design, and process and construction engineering. This demand will grow well into the year 2000. Increased demand in industrialized countries (where new compliance standards focusing on particulate emission are in place) will continue in the area of air pollution control equipment.

The World Health Organization (WHO), in cooperation with the United Nations Environmental Program (UNEP), monitors the Global Air Quality Monitoring Project (GEMS/Air). This project provides data on air quality trends in major cities and nations around the world. This data can be easily accessed on the worldwide web at: http://www.who.ch/peh/.

Growth and expansion of the international environmental market for the future will depend to a large extent on industrial economic growth and the trend in the development of pollution prevention control technologies concurrent with national and global enforcement of environmental regulations and standards—for example, the global ambient air quality standards under the Montreal Accord.

7.3.2 Global Water Quality and Sanitation Indicators and Trends

Human and environmental health are inextricably connected to global water quality. Although there have been tremendous national and global efforts to provide sanitation services and to reduce the extent of surface and groundwater contamination, global water quality has continued to deteriorate, largely in developing countries (Figures 7.4a and 7.4b). Contributing to the problem are untreated human sewage, industrial effluents and seepage of commercial chemical byproduct, improper management and disposal of hazardous wastes

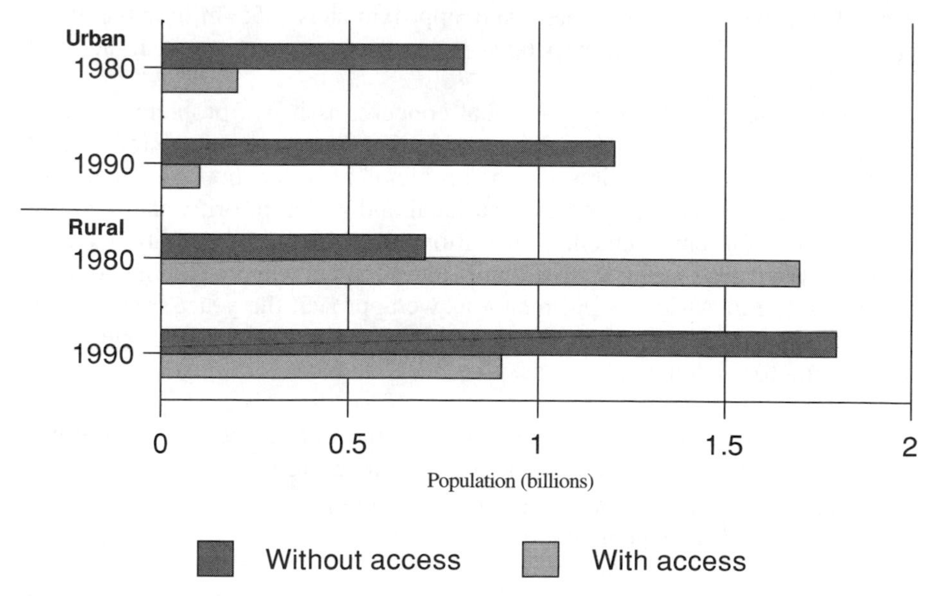

Figure 7.4a Access to safe water in developing countries, 1980 and 1990 (*Source:* World Health Organization)

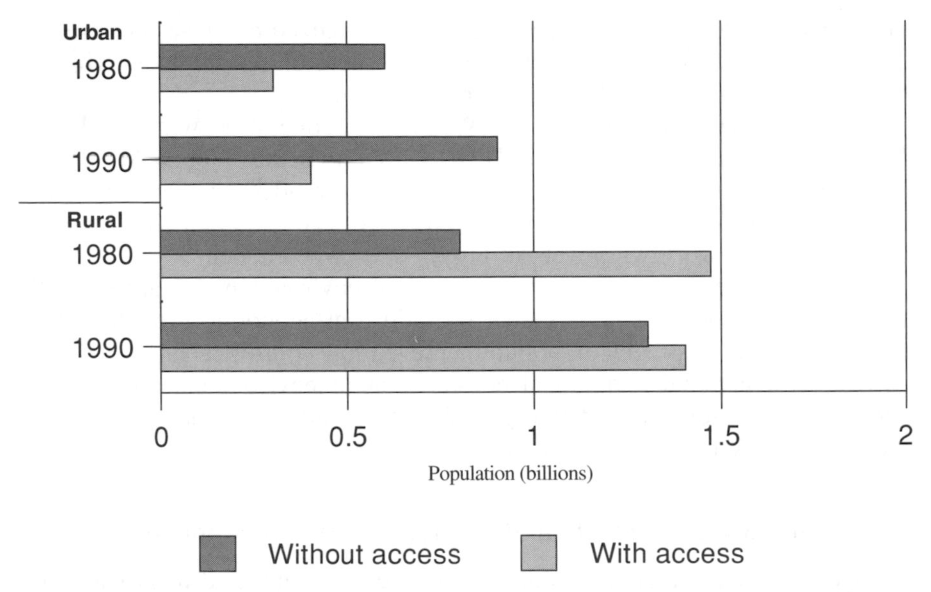

Figure 7.4b Access to adequate sanitation in developing countries, 1980 and 1990 (*Source:* World Health Organization)

and uncontrolled, widespread application of fertilizers, herbicides, and pesticides. There are also natural contributing factors such as droughts and decrease in seasonal rainfall, but the predominate contributing factor is population growth. As shown in Figure 7.4a, in 1990 about 1.2 billion people in urban areas and 1.7 billion people in rural areas in developing countries lacked adequate drinking water, and approximately 855 million people in urban areas and 1.3 billion people in rural areas were without adequate sanitation.

Water scarcity is becoming a global concern, as it is a problem that faces every nation. Figures 7.5 and 7.6 provide data on water usage and scarcity within each continent. Addressing the problems of water shortage, water quality, and management has become a national and global priority, and the growing demand for environmental practitioners within the international arena (especially in arid regions such as the Middle East and in developing countries such as those in Africa) is promising as we approach the year 2000. In many developed countries the need for consulting services related to water issues is growing for two major reasons:

1. The shift in water management from the public sector to the private sector and the resulting marketplace competition
2. The increasing demand for more effective/efficient technology application and management

Likewise, in developing countries where population growth and drought are outpacing infrastructure development, concern about water issues is growing

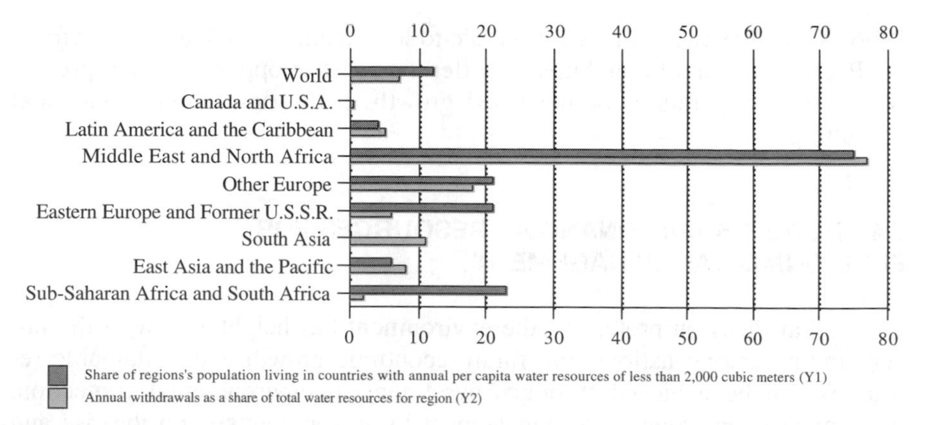

Figure 7.5 Global water use and scarcity—1970s, 1980s, and 1990s data (*Source:* World Bank Environmental Data)

quickly as demand overshadows supply. The lack of technology and expertise in developing countries has, in part, increased the market opportunities for foreign environmental consultants and companies. The major growth has been recorded for specialty chemical companies of the United States, which alone make up the largest portion of the international market (approximately 40 percent of international revenues as reported by the *Environmental Business Journal* (*EBJ*), February 1996). Water management equipment and related technologies are in high demand, which will continue to grow well into the twenty-first century. America's top 50 water technology manufacturing management companies reported a gain in international sales from 22 percent in 1991 to 30 percent in 1993 to 31 percent in 1995 (reported by *EBJ,* February

	Annual Internal Renewable Water Resources		Percentage of Population Living in Countries with Scarce Annual per Capita Resources	
Region	*Total (thousand of cubic kilometers)*	*Per Capita (thousands of cubic meters)*	*Less than 1,000 Cubic Meters*	*1,000–2,000 Cubic Meters*
Sub-Saharan Africa	3.8	7.1	8	16
East Asia and the Pacific Rim	9.3	5.3	1	6
South Asia	4.9	4.2	0	0
Eastern Europe and the former USSR	4.7	11.4	3	19
Rest of Europe	2.0	4.6	6	15
Middle East and North Africa	0.3	1.0	53	18
Latin America and the Caribbean	10.6	23.9	1	4
Canada and the United States	5.4	19.4	0	0
World	40.9	7.7	4	8

Source: World Resources Institute Data, World Resources Institute 1992 Report World Bank Data, 1992.

Figure 7.6 Water availability by regions

1996). Countries in regions with middle to low incomes, such as Asia, Africa, the Pacific Rim, and Latin America, offer the greatest opportunities at present and a promising future for increased growth in the field of environmental consulting.

7.4 BUDGETS AND FINANCIAL RESOURCES FOR ENVIRONMENTAL MANAGEMENT

The global focus on protecting the environment has heightened with the understanding among nations that future economic growth and sustainable resources can be achieved through investment in environmental protection. Investment in environmental management by governments is on the rise and it is most likely to further increase as we approach the year 2000. The ability of industrialized countries to benefit economically and achieve a maximum sustainable future depends on access to trade and markets, which in turn depends on the protection of natural resources and the environment. Therefore, industrialized countries continue to increase their capital investment in environmental protection and management. These countries are instituting optimal environmental policies and programs that balance the initial high cost of environmental management with the benefits of ensuring potential economic gains through increases in productivity, human health, human survival, soil conservation, and other environmental sectors.

Developing countries have now begun to institute environmental policies through the implementation of National Environmental Action Plans (NEAPs), especially in Africa. Other developing countries have established environmental agencies to create and implement regulations, requirements, and laws to address environmental issues. This trend is likely to grow beyond the year 2000 in developing countries. The realization that investment in environmental programs, especially in developing countries, will yield steady economic growth and stability has lead to an increase in private financing by international financial institutions and private institutions. Table 7.4 presents an example of short-term costs to environmental programs in developing countries and their associated long-term benefits.

7.4.1 Where Does Public and Private Environmental Financing Go?

Most international, national, and state governments have instituted policies that provide incentives for industries and businesses to provide economic growth and reduce negative impacts on the economy. The combination of governmental actions, environmental policies, and the convergence of private and social interests for sound environmental practices generate investments in technical and managerial environmental experts in areas such as the following:

- Education and training

Table 7.4 Estimated Costs and Long-Term Benefits of Selected Environmental Programs in Developing Countries

Programs	*Additional Investment in 2000*			Long-Term Benefit
	Billions of dollars a year	As a percentage of GDP in 2000[a]	As a percentage of GDP growth, 1990–2000[a]	
Increased Investment in Water And Sanitation	10.0	0.2	0.5	Over 2 billion more people provided W it service. Major labor savings and health and productivity benefits. Child mortality reduced by more than 3 billion a year.
Controlling particulate matter (PM) emissions from coal-fired power stations	2.0	0.04	0.1	PM emissions virtually eliminated. Large reductions in respiratory illnesses and acid deposition, and improvements in amenity.
Reducing acid deposition from new coal-fired stations	5.0	0.1	0.25	—
Changing to unleaded fuels; controls on the main pollutants from vehicles	10.0	0.2	0.5	Elimination of pollution from lead. More than 90% reductions in other pollutants, with improvements in health and amenity.
Reducing emissions, effluents, and wastes from industry	10.0–5.0	0.2–0.3	0.5–0.7	Appreciable reductions in levels Of ambient pollution, and improvements in health and amenity despite rapid industrial growth, low-waste process often a source of cost savings for industry.
Soil Conservation and afforestation, including extension and training	15.0–20.0	0.3–0.4	0.7–1.0	Improvements in yields and productivity of agriculture and forests, which increase the economic returns to investments. Lower pressures on natural forest. All sustainable forms of cultivation and pasture.

(*continued on next page*)

Table 7.4 (*Continued*)

Programs	*Additional Investment in 2000*			Long-Term Benefit
	Billions of dollars a year	As a percentage of GDP in 2000[a]	As a percentage of GDP growth, 1990–2000[a]	
Additional resources for agricultural and forestry research, in relations to projected levels, and for resource surveys	5.0	0.1	0.2	—
Increasing primary and secondary education for girls	2.5	0.05	0.1	Primary education for girls extended to 25 million more girls, and secondary education to 21 million more. Discrimination in education substantially reduced.

Source: World Bank, *World Development Report 1992* (Washington DC: Oxford Press), 1992.

Note: GDP of developing countries in 1990 was $3.4 million and projected to reach $5.4 trillion by 2000. Projected growth rate of GDP is 4.7% per year.

- Research, development, and technology demonstrations
- Environmental assessments and monitoring
- Corporate/business/industry environmental policies
- Community and public environmental awareness programs

The European Commission (EC) under its various Policies and Treaties have in place public financing that focus on environment, rural development, research and other areas. There are other programs that facilitate public financing for environment management.

7.5 ENVIRONMENTAL CAPACITY-BUILDING NEEDS IN DEVELOPING COUNTRIES

Developing countries are lagging behind developed countries in environmental management. Capacity building is defined as the utilization of the "state-of-the-art" technology transfer to strengthen a sector from the ground floor up. The link between environmental, social, and economic development and between poverty and environmental degradation are prominent in developing countries. These dynamic issues that confront developing countries have had a low priority in most of these countries, resulting in fewer resources and less capacity to manage the economy and the environment.

The Global Conference on Environment and Development held in Rio de Janeiro in 1992 set the agenda for an international consensus to address the issues of poverty and environmental degradation. Agenda 21 includes, among other items, a commitment by developed countries and the international donor community to assist developing countries in environmental capacity building.

Members of the international donor community, such as the World Bank, United States Agency International Development (USAID), (UNDP), and UNEP, are increasing their efforts in providing environmental capacity building in developing countries. Other international organizations, such as the African Capacity Building Foundation and other nongovernmental organizations, have also increased their activities toward providing environmental capacity building.

The following are areas and sectors in which capacity building is essential in developing countries:

Technical Aspects

- Development of solid waste management and hazardous waste experts
- Provisions of resources for management and application of disposal, treatment, transport, and storage technologies
- Development of environmental technical training facilitators to teach technical design approaches

Socioeconomic Aspects

- Development of capacities to ensure increased budgets and financial resources for governmental agencies involved in fiscal management and infrastructure development
- Streamlining local government and central government administrative practices and programs to achieve the optimal objectives of the government in providing services

Environmental Regulation/Enforcement Capacity and Policy

- Strengthening of environmental requirements and regulations
- Strengthening of compliance oversight and enforcement capacity
- Strengthening of environmental planning and management at both the municipal government level and the central government level
- Strengthening of environmental management practices in industries and businesses

Service Sector Growth

Service sector activities are defined as environmental projects related to services at any stage of an environmental activity and/or management. These are usually activities that entail:

- Data gathering and fact-finding
- Preliminary analysis
- Preliminary feasibility study and audit
- Project design
- Evaluation and review

This sector had tremendous growth a decade ago in developed countries during the emergence of environmental regulations, policies, and standards. The service sector boom in developed countries slowed in the 1990s, when most of the regulated universe and potential environmental clients matured in their knowledge and practices of sound environmental management in accordance with established rules, regulations, and the best applicable technologies and equipments.

In the developing countries, such as those of the Pacific Rim, parts of Asia (such as Indonesia), and Africa, the emergence of environmental management as an integral part of economic and social development has increased the demand in the environmental service sector. This trend will continue to grow well beyond the year 2000.

Management Sector

The management sector includes on-site environmental management activities such as:

- Process monitoring, record keeping, and tracking
- Supervision and control
- Training
- Facilitation

Demand for consultation services within the management sector in developing countries maintains marginal growth, owing to the growth of the regulated environmental universe, creation of environmental departments, and hiring of corporate environmental managers. Growth of the management sector in developed countries will likely be constrained, as most companies may see only a slight increase in demand due to the increase in new environmental legislation, amendments to existing regulations and standards, and the globalization of environmental standards.

Construction Sector Growth

The continuing development of pollution prevention technologies and pressure to incorporate cost-effective techniques within the manufacturing and the industrial sectors will lead businesses to acquire new technologies and to construct and retrofit their industrial processes to eliminate and/or control industrial wastes and reduce environmental risk. Industrial waste generators and governmental environmental regulators have come to the conclusion that it is a commonsense approach as well as a preferable, technically sound approach, to address environmental problems by preventing pollution in the beginning, instead of addressing environmental problems at the "end of the pile." The environmental construction sector will, therefore, continue to experience tremendous growth as information technology, environmental technologies, and telecommunication advances become available globally, especially in developing countries. The environmental remediation sector will continue to boom, as leading to greater demand in the environmental construction sector as remedial design and construction at polluted sites and facilities increase. Environmental clean up and corrective action at active facilities and industries will become prevalent as environmental safety and health become high priorities in most countries with low-income levels, high poverty rates, and high mortality. The opposite effect will be evident in developed countries, such as the United States where the Superfund Program is nearing its full maturity and will level off as we approach the year 2000. By the later part of this century growth in the developed countries will be slow and will commence to drop in the beginning of the next century. Therefore, the best opportunities and markets for construction are in developing countries.

7.6 ENVIRONMENTAL TECHNOLOGIES AND EXPORT MARKET/ TECHNOLOGY TRANSFER

Successful environmental investments include appropriate technologies to address various environmental problems. The trend in the environmental technologies export market has shown rising competition between developed and industrialized countries as many developing countries invest in environmental management and natural resource protection. In addition, the increasing public demand in developed countries for stringent environmental regulations has led businesses and industries to research and develop new technologies to minimize the cost of maintaining environmental management and abiding by those regulations. The environmental technologies export market is likely to maintain its growth as we approach the year 2000, and exports will increase with the increase in environmental priority in developing countries. This trend will follow an increased demand for environmental consulting and experts from leading industrialized countries with companies that have developed environmental technologies. The United States leads the world in environmental technologies, especially in "green technologies," followed by Germany and Japan (each has strong standing in water treatment technologies). Opportunities for United States manufacturers and consulting companies are increasing, because more of the new technologies developed in the United States are commercially available and proven. The market demand will continue to be driven by in-country government environmental policies, demand from the public for a safer environment, global environmental treaties and agreements, and the growing need for environmental goods and services to meet regulatory standards and to control pollution before it occurs. Demand is strongest in developing countries in Africa and middle-income country markets such as Brazil, Thailand, Indonesia, and Mexico.

Globalization of the environmental management industry is maturing at a rapid rate. The opportunities and growth today lie in developing countries with middle income levels and will strengthen in lower-income developing countries as we approach the year 2000. Environmental technology transfer and pollution control technology markets present the greatest opportunities in developing countries. The promising areas for rapid growth are the following:

- Water treatment equipment, technology, and chemicals
- Sanitation, sewage treatment, and water equipment technologies
- Resource recovery (such as renewable energy, waste-to-energy technology)
- Air pollution abatement
- Solid waste and hazardous waste management (e.g., landfill construction)
- Environmental policy and capacity building
- Environmental training and consulting
- Testing and analytical services, environmental impact assessment (EIA) and site remediations

8

PLANNING AND RESEARCH (BEFORE YOU STEP INTO A FOREIGN COUNTRY)

Now that you have acquired and equipped yourself with information about international environmental consulting practice, you are ready to make your first visit to the foreign country where you plan to do business.

The world economy and environment have been undergoing rapid changes and shifting toward an integrated system. Globalization of the world economy and environmental management has led to an increased rate of international business and consulting practice. Even so, exposure to the different cultural, political, and economic systems of the global market has not been given high priority by international business entities and practitioners. To be fully functional in a market, a consultant must be conversant with the different attributes and configurations of that market, the significant elements of which pertain to the culture of the country. This chapter discusses the things a consultant must do before entering a foreign country to do business.

8.1 THINGS TO DO

When you get to your destination, there are a number of things you should do:

- *Contacts.* Let your contacts and other business affiliates know of your arrival. Make sure you set an appointment with each contact person to go over your scheduled activities and ask for input. Make an arrangement with your contact person for the weeks' agenda.

- *Communication.* If you require a translator, make sure he or she is aware of your arrival and where you are staying.
- *Client Interaction.* If you are dealing with the government sector, be aware that it is very difficult at times to schedule appointments. Most government officials, especially in developing countries, are often unavailable; many are out, engaged in their personal business. In some cases you must learn the system of operation to enable you get through to the person(s) with whom you need to discuss your business issues. You may want to request your in-country contacts to do the "legwork" to ensure timely and appropriate responses to officials you intend to see.

  If you are dealing with the private sector, be sure to understand the other party's needs and wants. Within the private sector people want to see a quick turnaround and delivery of services and goods. In developing countries you will find that to any proposition to a businessperson or entity in the private sector, the usual answer is "No problem." In many developing countries it is customary for a businessperson to say everything is okay when it is not. The phase "no problem" is used often. Be careful and evaluate thoroughly any agreement in principle you make.
- *Customs and Business Ethics.* There may be several sets of traditions and customs within a single country. In many cases, because of tribal and ethnic diversity, you may be confronted with complex communication barriers in your business dealings; therefore, you should be sensitive to the business customs and know how to deal with diverse situations. Knowing the attributes of a culture will help you to gain the trust of the native businesspeople you are dealing with. You should become acquainted with local customs as fast as possible. A good translator can usually guide you and protect you against gross mishaps. It is also important to be alert to the business climate and ethics of the region.

  The following are a few questions you should ask (and seek answers to) when preparing to conduct business in a foreign country:

  How do the native people conduct business?

  How does the government bureaucracy function?

  How do you get influential people to pay attention to your needs?

  What is the chain of command in governmental institutions?

  How do companies operate?

  What are the most important goals and objectives of business entities?

  What are the do's and don'ts of the business culture (e.g., appropriate and inappropriate attire to wear to business meetings)?
- *Banking and Finance Operations.* In many countries you will be faced with government taxes on profits, interest, and other taxable income for foreigners doing business in those countries. The tax structure differs from that applicable to in-country native business entities. You should obtain information from a country's central bank or the appropriate min-

istry or government agency. This is very important, because failure to adhere to these laws may cause you to lose the license to practice and do business in that country. You could face fines and prohibition from doing business in the country.

8.2 ASSESSING KEY POLITICAL AND ECONOMIC SENSITIVITY FACTORS

The political and economic factors are specific elements a foreign consultant must be aware of in order to effectively function in a particular country. Understanding these elements entails assessing the factors external to your regular business and consulting practice that are directly and indirectly affected by the econo-politico-cultural norms specific to the country. The presence and activities of private interest groups, laws, branches of government, lobbyists, political parties, and businesses form a political climate that influences and shapes the economic climate of a country. Politics and economics are not neutral or coincidental to each other. Both deal with the issues of people, their social life and well-being, economic functions and activities. The functional elements of politico-economics concerned with the allocation of resources for the competing market forces involve various types of conflict between social classes and between political and business groups and individuals within classes. This framework of politics and economics is different in every country. Moreover, the political, economic, and business environments are dynamic.

The dynamics of these environments are not surprising, however, when one considers that there are external influences that shape and dictate each country's soicoeconomic and political structures. Thus, when operating in another country, you should respect and be sensitive to the governance of the nation, its regulations, requirements, and laws, its statehood/autonomy, the way the bureaucracy operates, and, finally, the mechanisms and cultural barriers that may obscure the daily operations of your business activities in that country. You have to assess these factors, as discussed earlier in relation to what to do before entering that country. Upon entering you must know these sensitive factors and how to deal with them to get your business activities to a successful conclusion; for example:

- To what extent do the political, economic, and cultural climates differ from those in your country?
- What are the general values, and what do they imply?
- What are the likely value conflicts you may encounter while doing business in another cultural setting?
- What are the basic government policies, regulations, standards, and requirements for a foreign business entity?

- What are the general economic systems, functions, and operations in the country?

Answers to these questions provide the basic tools to effectively function in an international setting. It is necessary to be aware of these issues if your international environmental practice is to be successful, bearing in mind that transfer of your traditions and cultural attributes to other cultures can make or break your success. It is also important to know that the culture and business practices of a country may be so diverse that you will find it very difficult, or impossible, to penetrate the market if you do not have knowledge of how things operate in that country. You must exhibit a professional ethical code of sensitivity to cultural diversity, understand the functional elements of the in-country economic systems, and know the governmental policies, regulations, requirements; moreover, you should familiarize yourself with the way business is conducted in that country. These essentials are highlighted by Black and Mendenhall (1991, p. 178), who noted that "negotiations between business people of different cultures often fail because of problems related to cross-cultural differences." Von Glinow and Teadarden (1988, in Mendenall and Oddu, 1991, p. 320), in discussing problems inherent in the transfer of human resource management technology, said, "HRM (Human Resource Management) technology will have to be modified to accommodate many of the Chinese system constraints." Harris and Moran (1991), writing about the globalization of management theory, stated:

> While global managers are open to management innovations from abroad . . . oversimplification can lead to dangerous assumptions, so international leaders need cultural sensitivity in their analysis of world literature and trends in management and commence. (p. 13)

White and Rhodeback's "Essay on Ethical Dilemmas in Organization Development: A Cross-Culture Analysis" (*Journal of Business Ethics,* 1992, p. 664) wrote:

> The field of organizational development cannot afford to be less vigilant than other disciplines in the pursuit of knowledge concerning the implications of multicultural similarities and differences for a successful international professional practice. Particularly critical is the need to test the assumption that ethical standards for professional conduct are transferable to other countries.

In fact, the complexity of integrating into the international market place cautions international businesses and consultants to accustom themselves to the broadest global culturalism, value systems, and in-country economic and political functional systems.

Understanding the importance of socioeconomic and political factors is critical to successful entrance into the international environmental consulting market. General orientation to simple but useful tools can provide the basic information necessary to deal with in-country economic, political, and cultural mores. Patterns and reforms of political, economic, and social attributes of a country have an impact on a foreign investor and on conducting business in that country. Macroeconomic stabilization, liberalization of trade and internal markets, and in-country institutional reforms are, among others, characteristics that may affect the smoothness of doing business in a country abroad. In any case, the consultant should develop a specific approach to obtain specific data that will affect his or her consulting practice or company's business activities in the country in which he or she intends to operate. There are a number of ways to obtain basic information quickly; some of these are listed in Table 8.1.

This following section discuss the general political, economic, and, in some cases, traditional cultural attributes within each continent in the world. Current data can be obtained from the U.S. State Department's National Trade Data Bank.

8.2.1 Africa

Africa is bordered on the north by the Mediterranean Sea, on the west by the Atlantic Ocean, and on the east by the Indian Ocean. The continent comprises 52 countries, spreads over an area of about 117 million square miles, and covers about 21 percent of the earth's surface.

The continent includes several tribes with different languages and cultures, providing the traveler with a unique experience. Transportation is far less efficient here than in the West.

Population. The population of Africa is about 712 million, estimated to a rise to about 3 billion by the year 2050. The present population is approximately 15 percent of the world's population.

Southern and Central Africa. The region consists of Angola, Botswana, Burundi, Central African Republic, Congo, Lesotho, Malawi, Mauritius, Namibia, Rwanda, South Africa, Swaziland, Uganda, Zaire, Zambia, and Zimbabwe. The total population of these countries is approximately 151 million,

Table 8.1 Information Sources for Economic and Political Factors Applicable to Doing Business in a Foreign Country

Sources
• Internet
• Country's embassy
• Library (information on country)
• Local or national organizations and associations affiliated with country
• U.S. Commerce Department
• U.S. State Department
• International business organizations and associations

with a total land are of about 3.4 million square miles. The region is bordered by the Atlantic Ocean and consists of desert and savanna lands, with interior plateaus of high elevations stretching across the central portions of the region. These high elevations are mostly arid and rocky formations. The climate for the most part is tropical, but dry and hot in the desert regions.

Natural resources predominant in the region are potash, diamonds, gold, oil, iron, cooper, salt, natural gas, timber, fish uranium, cobalt, chrome, bauxite, platinum, limestone, and coal. Environmentally protected areas are few. For example, protected areas in Angola consist of 1 percent of the country's area, in Botswana 18 percent, Central African Republic 6 percent, and Congo 4 percent.

East Africa. This region of Africa includes Comoros, Kenya, Djibouti, Ethiopia, Madagascar, Mozambique, Seychelles, Somalia, and Tanzania—a population of approximately 140 million with a surface land area of 1.9 million square miles. The area consists of highlands in countries like Kenya and Ethiopia. The climate in this region is generally hot and humid, but the highland areas tend to be cooler and moist, as compared with the low plains areas in the region. Major environmentally protected areas in Ethiopia consist of about 6 percent of the country's area in Kenya about 5 percent. Madagascar about 2 percent, Seychelles (unknown), Tanzania about 13 percent, Djibouti, Mozambique, and Somalia 0 percent, and Comoros less than 1 percent. Natural resources found in the region are gold, limestone, various minerals, salts, copper, platinum, chromium, graphite, semiprecious stones, uranium, and diamonds.

West Africa. This region, also known as sub-Saharan Africa, consists of Benin, Cameroon, Cape Verde, Côte d'Ivoire, Equatorial Guinea, Gabon, Gambia, Ghana, Guinea, Guinea-Bissau, Liberia, Nigeria, Sao Tome and Principe, Senegal, Sierra Leone, and Togo. The total population of this region is approximately 179 million, with a surface land area of approximately 1.2 million square miles. The region includes deciduous forest, forested plateaus, rain forest, and mangrove swamps around the coastal areas. Inland ares are of an arid, savanna-type topography. Climatic conditions such as humid climate condition with tropical, subtropical, semi-arid deserts and rains are harsh in many areas of the region.

North Africa. Algeria, Burkina Faso, Chad, Egypt, Libya, Mali, Mauritania, Morocco, Niger, Sudan, and Tunisia make up the region of North Africa. Its total population is approximately 364 million, and its surface land area approximately 8.1 million square miles. The topography of the region is includes the Atlas Sahien mountains in the north, the Sahara Desert in the south, and many areas of lowlands, plateaus, and savanna. The climate is hot and dry, with mild temperatures along the coastal areas. Major natural resources are crude oil, natural gas, iron, gypsum, uranium, and manganese.

Political Conditions. Most countries in the region have a democratic system of governance, but the Arab nations have a hierarchical system. Military rule is minimal in the region.

Economic Indicators. Most of the countries in the region are classified by the World Bank as low-income economies, except Algeria, Morocco, Tunisia, Libya, Côte d'Ivoire, Senegal, Cameroon, Gabon, Congo, Namibia, Botswana, Lesotho, Angola, South Africa, and Swaziland, which are classified as middle-income economies. Low-income economies are those with a gross national product (GNP) per capita of $675 or less in 1992, and middle-income economies are those with $676 to $8,355 per capita. Table 8.2 gives some basic socioeconomic indicators for the region, and country-specific basic economic indicators are included in Appendix XVII of this book.

Environmental Conditions. Weak governmental institutions, lack of technical expertise, and lack of enforceable regulations allow polluters to dispose of and mismanage their waste. Environmental degradation and deterioration are on the rise because of the low priority given to the environment by regulatory authorities and the government at large and the lack of technology to manage waste generated from "cradle to grave." Table 8.3 summaries the environmental conditions in the various regions of Africa.

Business Organizations. In most African countries, foreign entities interested in doing business are permitted to enter any chosen field of practice. In the environmental consulting field, there are no restrictions, but you must double-check with the appropriate in-country governmental ministry or department. Environmental consultants, environmental technologies/products manufacturers, and companies are allowed in many countries within the continent to establish limited liability companies, partnerships, and joint ventures.

Trade. Registration and licensing may be required in many countries. In most cases, individual consultants are not required to be registered or licensed

Table 8.2 Past and Present Socioeconomic Indicators in Africa

Indicator	Historical Africa	Current Africa
GNP per capita	$340 (1980)	$340
Life Expectancy	48 yrs (1982)	52 yrs
Infant Mortality	142/1000 (1970)	99/1000
Child Mortality	N/A	170/1000
Adult Literacy	27% (1980)	50%
Primary School Enrollment	50% (1970)	66%
Secondary School Enrollment	7% (1970)	18%
Annual Population Growth	3.0% (1980–1985)	4.2%

Source: United States Agency for Internation Development (USAID).

Table 8.3 Environmental Condition Ratings in Africa

Region	Rating	Score
Southern and Central Africa	The horrors of war, famine, drought, and environmental destruction are inescapable in this region. Hungry people do not have the luxury of thinking about the future. Outside aid is desperately needed before it is too late.	3
East Africa	Although the region has some good environmental polices, overpopulation, civil unrest, and drought have overburdened this once-lush area. If population growth can be limited, the area may be able to return to its natural beauty and abundance.	3
Sub-Saharan Africa	Nationa here pay lip service to conservation.	4, 5
North Africa	North Africa has been depleted by drought, hunger, and civil war. As people leave the plundered and unproductive countryside for the cities, they find even less chance of making a living. This region pays lip service to the environment, but little action is being taken.	4

Key to Scoring

1 Nations in the region are undertaking conservation initiatives of all types and have good policies in place.

2 Nations in the region are beginning to undertake conservation initiatives, but are threatened by traditonal and cultural habits, therefore, they need more encouragement and/or assistance to achieve their conservation goals.

3 Nations in the region may have joined international initiatives and have followed up with some actions, but the region is in danger and much more needs to be done to aid or improve conservation.

4 Nations have shown a green front, that is natural resource preservation, but mainly pay lip service to the environment, with little or no action.

5 Environmental impacts are severe, and nations in the region are not actively responding with conservation initiatives.

Peter A. Sam, 1996.

if the project is a short-term, one-time activity. Once again, you may have to double-check with the appropriate in-country ministry for information. There may be import and export restrictions, and you should contact local officials in-country about specific provisions for trade and commerce.

8.2.2 Europe

Environmental conditions in Europe vary from one location to another; Table 8.4 presents fair summaries of conditions in the regions of Europe. Specific regional issues are elaborated further in the following sections of this chapter.

Table 8.4 Environmental Condition Ratings in Eastern and Western Europe

Region	Rating	Score
Western Europe	Western Europe has good environmental polices—and big environmental problems. This is an uphill battle, but there is hope that the region is heading uphill, not down.	1, 2
Eastern Europe	Eastern Europe's people are suffering the consequences of industrialization without environmental protection. People are working hard with what little they have, but the problems are great.	3, 5
Former Soviet Union	The area is in severe danger. Local governments are having trouble organizing environmental efforts in increasingly nationalistic states. Yet the people have overcome political suppression and may yet overcome seemingly insurmountable environmental problems.	3

Key to Scoring

1 Nations in the region are undertaking conservation initiatives of all types and have good policies in place.

2 Nations in the region are beginning to undertake conservation initiatives, but are threatened; they need more encouragement and/or assistance to achieve their conservation goals.

3 Nations in the region may have joined international initiatives and have followed up with some actions, but the region is in danger and much more needs to be done to aid or improve conservation.

4 Nations have shown a green front, that is natural resource preservation, but mainly pay lip service to the environment, with little or no action.

5 Environmental impacts are severe, and nations in the region are not actively responding with conservation initiatives.

Source: Earth Journal, 1993.

Western Europe. The region consist of Andorra, Austria, Belgium, Denmark, Finland, France, Germany, Greece, Iceland, Ireland, Italy, Liechtenstein, Luxembourg, Malta, Monaco, the Netherlands, Norway, Portugal, San Marino, Spain, Sweden, Switzerland, the United Kingdom, and Vatican City. The region's population is approximately 378 million, with a surface land area of approximately 2.3 million square miles.

The Alps connect Austria, France, Germany, Switzerland, Italy, and Liechtenstein. Geographically, the plains along the central and southern parts of the region are rich and very productive. There are hills and rivers between the mountain ranges. The climate of the northern region is predominantly cold, coverings with icy conditions and snow. Major natural resources of the region include tungsten, pyrite, petroleum, mercury, copper, dolomite, gypsum, silver, peat, barite, lead, coal, iron, timber, potash, marble, salt, and timber.

Political Conditions. The present political conditions are stable for foreign environmental consulting practice. However, in the near future the European integration process could affect the political climate of the region. The Maastricht Treaty and the ongoing process of European union, which includes the integration of a common European currency and security policy, have tremendous political ramifications. There will be a shift to a much more centralized governmental system, with member countries still maintaining autonomy. The political integration will definitely have an impact on trade and business. At present, the plans for integration include creating a single market economy. Therefore, new political avenues will be created that will influence the decision-making processes. Thus, environmental policies will certainly take a major turn as many environmental regulations will have a direct impact on trade and the economy, such as in the areas of consumer product standards and the transboundary movement of toxic waste and hazardous constituents.

Economic Indicators. The countries in the region are classified as middle- to high-income economies with a per capita GNP of $675 to $8,355 and higher. The region is expected to maintain economic growth for the rest of the century. The ongoing process of European Union economic integration will increase free trade and provide increased flexibility in doing business.

Business Organizations. There are several professional business organizations in Europe that offer assistance to their members in trade and business practices. There are no restrictions on environmental consulting practice for either resident or foreign practitioners. Registration and licensing of a business is required in most of the countries in the region. Nevertheless, there may be waivers for a short-term consulting practice offered by a foreign entity.

Trade. Registration and licensing may be required in many countries in the region. In most cases individual consultants are not required to be registered or licensed if the project is a short-term, one-time activity and does not require a professional practicing license. You may have to double-check with the appropriate in-country governmental department/agency or request from the contracting office the information needed in meeting any licensing requirements. There may be import and export restrictions, and you should contact local officials in-country in regard to specific provisions for trade and commence.

Environmental Condition. The region consists of countries that are developed and industrialized; therefore, they face the general environmental problems that predominate in industrialized countries such as the United States. The most important environmental problem stems from the geographic location and closeness of the countries within the region—that is, its transboundary air pollution problem. Activities of the coal industry have been major sources of acid rain in the region. High levels of emissions of particulates such as sulfur dioxide have affected the forests. Other environmental problems lie in the area of environmental management and restoration. Past practices in hazardous waste disposal have resulted in many contaminated sites throughout the region. For example, in West Germany there are approximately 50,000 dump sites awaiting cleanup. Urbanization and population growth in urban centers have led to increased waste generation, also a major environmental problem facing the region. Waste minimization efforts, such as recycling and waste-to-energy initiatives, can alleviate this problem.

Eastern and Central Europe. This region includes Albania, Slovenia, Romania, Hungary, Yugoslavia, Czechoslovakia, Croatia, Bulgaria, Poland, and Bosnia-Herzegovina. The population of the region is approximately 123.7 million, with a surface land area of about 450,000 square miles. About 50 percent of the area is agricultural land and the rest is plains and forest lands. The climate is cold in winter and slightly warm in summer, with occasional hot weather. The natural resources of the region consist of timber, oil, minerals, metals, coal, ignite, and natural gas. The environmentally protected areas include 2 percent of the area in Albania, 1 percent in Bulgaria, and 16 percent in Czechoslovakia.

Political Conditions. The collapse of the Soviet Union and the end of the Cold War have led to political reform and transformation of the region's politics. Poland, Slovenia, Yugoslavia, Croatia, Bosnia-Herzegovina, and Albania, to mention a few, have commenced political reforms that are expected to led to further reforms, such as economic and geographical changes. The new direction toward establishing democratic institutions provides a harmonious setting for international business in many of the countries in the region. The political changes in the countries of Central and Eastern Europe countries will continue to shape the economic conditions in the region.

Economic Indicators. Across the region, there are signs of economic growth with real GDP expected to increase, as compared with the sluggish economy of the late 1980s and during the civil war in many parts of the region. There are already promising signs of growth in the most developed/industrialized countries, such as Poland, the Czech Republic, and Hungary.

Business Organizations. Privatization has opened the doors to foreign investment and foreign business. The shift to profit-oriented business activities has increased access to the environmental market. The once nationally owned businesses are now privately owned, making the process of engaging in business less bureaucratic. Privatization has advanced in the Czech Republic but has stalled in Hungary and Poland. The significant improvements resulting from privatization have led to other reforms, such as industrial restructuring and trade liberalization, especially in the Czech Republic. This regional business climate is a fruitful ground for foreign environmental investors and businesses.

Trade. The structural adjustment program and the privatization efforts being undertaken in many countries in the region illustrate the new direction of growth-oriented economic policies, which are likely to spur the growth of environmental markets. Funding for environmental projects is largely acquired from international multilateral organizations and international development aid agencies. Registration and licensing to do business are required in some cases. Most countries in the region have a bilateral trade treaty and are classified and considered as Most Friendly Nation (MFN).

Environmental Conditions. The past neglect of environmental protection, lack of adequate environmental policy, and lack of laws and regulations have led to environmental deterioration and degradation of the ecology of the region. The predominant industrial activities, especially coal mining, have not paid much attention to pollution control, which has resulted in soil, surface water, groundwater, and air contamination. High levels of sulfur dioxide in the atmosphere have been recorded in some parts of the region, exceeding the allowable safe limits by as much as 50 percent, owing to the burning of coal. For example, in Katowice in the Upper Cellose, a major mining region in Poland, lack of appropriate pollution control technologies and expertise have compounded environmental problems. The industrial residue and pollutants from Soviet military activities in some parts of the region have resulted in the contamination of environmental media, mostly by hydrocarbons. Civil wars in the region, such those in Yugoslavia and Poland, have resulted in the contamination of the environment and destruction of habitat, loss of biodiversity, deforestation, and loss of soil fertility.

Former Soviet Union. The region includes Armenia, Azerbaijan, Belarus, Georgia, Kazakhstan, Kyrgyz Republic, Moldova, Russia, Tajikistan, Turkmenistan, Ukraine, Uzbekistan, Estonia, Lithuania, and Latvia. The region's population is about 427 million, with a total surface area of about 8.6 million square miles. The climate is varied: cold, with arctic conditions in the north, humid in the south, and cold, dry weather intermingling across the region. Most of the region consists of plains. The Frozen tundra is found in the northern section of the region. The central region consists of forest and vast grasslands. Natural resources include nickel, potash, phosphate, timber, mercury, oil, and lead. Environmentally protected areas comprise 1 percent of the region.

Political Conditions. This region is undergoing political and economic reforms, changing from a socialist system to an open, democratic system of governance. Russia, an industrialized-agrarian republic, is slowly going through a drastic sociopolitical and economic reform. The denationalization of the national corporation is opening the economy to a free market system. The monopoly of governmental agencies and systems has been reversed, and Western approaches in economic activities and functions are positive attributes of the current politico-economic system.

Economic Indicators. Overall, there seems to be a positive future in the economic environment for foreign investment and, doing business in the region. The economic factors are largely becoming similar to those of Western Countries as democratic reforms lead to economic readjustment toward open market systems. Privatization initiatives within the region show promising signs of economic growth.

Business Organizations. Business and trade organizations have sprung up in recent years as a result of economic reform, reorientation, and the shift toward privatization and open markets.

Trade. The political reforms in the region have had a drastic impact on economic activities. The new "openness" approach has led to increased entrepreneurship and decentralization. Adoption of the Western system offers much better conditions, as compared with the past Communist system, doing business. The region is expected to have a fruitful business climate in the future and to increase invitations to Westerners to provide technical assistance in many sectors, including natural resource management and environmental restoration.

Environmental Conditions. Lack of adequate environmental laws, regulations, and standards has left a legacy of contaminated soil, surface water, air, and groundwater owing to improper handling and disposal of industrial waste. Nuclear weapons and power plants, such as the Chernobyl nuclear power plant, have been the sources of radioactive releases resulting from improper management. Topsoil has been lost through erosion and, in many areas, is unsustainable for agricultural practices. For example, the Black Soil Belt region in the western areas have suffered lost of topsoil fertility because of the uncontrolled use of pesticides, fungicides, and fertilizers. The major environmental problems are contaminated hazardous waste sites, air pollution from industrial sources, and polluted rivers, lakes, and surface waters such as those of Lake Baikal and Lake Karachay.

8.2.3 Asia and the Pacific Rim

This region differs from one location to another; Table 8.5 presents fair summaries of the environmental conditions in Asia and the Pacific Rim. Specific regional issues are elaborated further in the following sections.

South Asia. The region consists of Afghanistan, Bangladesh, Bhutan, India, Maldives, Nepal, Pakistan, and Sri Lanka. It has approximately 1.15 billion people and covers about 2.1 million square miles of land. The geographic location is east of the Middle East. To the north are the Himalayan Mountains. The weather is extremely hot in the arid parts, rainy and humid in the mountainous parts. The natural resources of the region consist of oil, coal, iron, sulfur, zinc, bauxite, graphite, limestone, and precious stones. The environmentally protected areas consist of less than 1 percent in Afghanistan, 1 percent in Bangladesh, 19 percent in Bhutan, 4 percent in India, 7 percent in Nepal, 4 percent in Maldives, and approximately 10 percent in Pakistan and Sri Lanka.

Political Conditions. Regional political conditions are not totally stable. The Indian and Pakistani religious conflicts persist, and the standoff in respect to nuclear weapons is fragile. Afghanistan's internal reforms, initiated since the collapse of the Soviet Union, are promising. Despite a few political

Table 8.5 Environmental Condition Ratings in Asia and the Pacific Rim

Region	Rating	Score
South Asia	The area is threatened with population explosion and limited resources, yet local actions are beginning to empower local citizens.	2, 3
East Asia	This area is beginning to pay more attention to environmental protection; time will tell whether it will be lip service or genuine commitment.	4
Southeast Asia	Southeast Asia has not yet put the enviroment on its political agenda. The population puts enormous pressure on forests and other fragile environments. Enviromental groups here are just beginning to make an impact.	3
South Pacific	Low populations and sustainable practices make this area a bright spot on the environmental map. Australia is one of the countries leading the way in environmental policy.	1

Key to Scoring

1 Nations in the region are undertaking conservation initiatives of all types and have good policies in place.
2 Nations in the region are beginning to undertake conservation initiatives, but are threatened; they need more encouragement and/or assistance to achieve their conservation goals.
3 Nations in the region may have joined international initiatives and have followed up with some actions, but the region is in danger and much more needs to be done to aid or improve conservation.
4 Nations have shown a green front, but pay mainly lip service to the environment, with little or no action.
5 Environmental impacts are severe, and nations in the region are not actively responding with conservation initiatives.

Source: Earth Journal, 1993.

glitches, the existing political condition in the region is not a threat to foreign investors and businesses.

Economic Indicators. Countries of this regional are characterized as low-income economies with a GNP per capita of no more than $635.

Business Organizations. Business and trade activities within the region are well organized and operate at levels of efficiency close to those in developed countries.

Trade. The region exports gems, jewelry, engineering goods, chemicals, leather manufacturers, and cotton yarn and fabric. Imports are crude oil, petroleum products, and fertilizer. Some countries are accused of exploiting child labor.

Environmental Conditions. There are about five typhoons per year along the eastern and southern coasts. Air pollution results from excessive use of high-sulfur products and coal as fuel. Other environmental problems are water shortage, deforestation, and soil erosion, Countries in the area are signatories to many international environmental treaties covering whaling, wetlands, and similar subjects.

East Asia. This region includes China, Hong Kong, Japan, Mongolia, North Korea, South Korea, and Taiwan. The total population of the region is approximately 1.3 billion, with a land area of about 4.4 million square miles. The high mountains of the Himalayas dominate much of the region. Much of the fertile plains is used for agricultural. The climate ranges from humid to semitropical. Natural resources in the region consist of timber, natural gas, marble, metals, limestone, and coal. Protected environmental areas include 1 percent in China, 6 percent in Japan and South Korea, less than 1 percent in Mongolia and North Korea, 0 percent in Taiwan and Hong Kong.

Political Conditions. Except for recent threats from North Korea, political conditions are fairly stable. The region is predominantly democratic, except for the Chinese socialist government. This political climate does not impinge on the ability to conduct business successfully. The region is shifting from a semiprotectionist stance and is rapidly opening its markets to the West.

Economic Indicators. The economies of the region fall in the middle-income to high-income categories, with GNP per capita at $636 to $7,910 and higher. Economic growth has soared in the 1990s, and the region offers much better economic and trade conditions for foreign businesses and investors. Japan, the largest economy in the region, maintains a high growth rate (GDP).

Business Organizations. The organizations of South Korea, Japan, Hong Kong, and Taiwan are by far the most advanced in Western business practices. Others are moving away from communist-style command economies.

Trade. China's 1994 exports came to $121 billion; imports to $115 billion. Its external debt was $100 billion. Agriculture accounts for 30 percent of GDP. South Korea, Japan, and Taiwan have capitalist economies, as does Hong Kong, which, for decades, had been a British colony. Exports include textiles, toys, footwear; imports include rolled steel, automobiles, oil products, and other goods.

Environmental Conditions. Urbanization and population explosion continue to cause environmental deterioration, including air pollution resulting from excessive use of high-sulfur products and coal as fuel. Other environmental problems are water shortage, inadequate management of municipal solid waste, deforestation, and soil erosion. Countries in the area are signatories to many international environmental treaties covering whaling, wetlands, and other subjects.

Southeast Asia. The region comprises Vietnam, Thailand, Burma, Malaysia, Laos, Indonesia, Brunei, and Cambodia. The population is approximately 458 million, with a land area of about 1.7 million square miles. This region is characterized by tropical, rainy, and humid subtropical climates with extensive rain forests that covers about 75 percent of the area. Natural resources consist mainly of timber, precious stones, oil, natural gas, rubber, coal, nickel, cobalt, gold copper, tin, bauxite fluorite, and tungsten, Environmentally protected areas comprise about 8 percent of the land of Indonesia, less than 1 percent in Laos and Burma, about 3 percent in Malaysia and Vietnam, about 2 percent in the Philippines, about 9 percent in Thailand, about 4 percent Singapore, and 0 percent in Brunei and Cambodia.

Political Conditions. Most governments are stable and democratic. Some nations have adopted the political dispensation of erstwhile colonial Britain, regarding the British monarch as head of state.

Business Organizations. The South Pacific Bureau for Economic Cooperation was established on April 17, 1973. In 1988 it was renamed the South Pacific Forum. Its address is Ratu Sukuna Roa, GDP Box 856, Suva, Fiji. This organization is a good source of information on business and trade activities in the region.

South Pacific. The region includes Australia, New Zealand, Papua New Guinea, the Solomon Islands, Fiji, Kiribati, Tonga, Tuvalu, Vanuatu, and Western Samoa. The population is approximately 26 million, and the land area comprises about 3.3 million square miles. The region is comprised of about 590 islands. The climate ranges from tropical and rainy to dry to humid subtropical. Natural resources predominant in the region include bauxite, fish, silver, timber, gold, nickel, uranium, copper, iron ore, zinc, lead, tin, oil, and coal. Environmentally protected include 5 percent in Australia, 11 percent in New Zealand, less than 1 percent in Fiji, Papua New Guinea, and the Solomon Islands, and 0 percent in the other nations in the region.

Political Conditions. This region enjoys a considerably high level of political stability.

Economic Indicators. The countries of Southeast Asia have low-income economies and low GNP, except for Australia, which has a higher purchasing power and a higher GNP.

Prosperous Australia has a Western-style capitalist economy. Its per capita GDP is comparable to that of industrialized Western Europe. In 1994 its exports amounted to $50 billion, and its imports to $51 billion. New Zealand's figures were $11 billion for exports, $10 billion for imports. Australia's inflation rate was 2.5 percent; New Zealand's, 1.6 percent.

Business Organizations. Many business organizations have been formed to promote commercial activities at home and abroad. For example, the Pacific Basin Economic Council, formed in 1967, links and advises business and government leaders interested in the region's business activities and trade opportunities.

Trade. The region's biggest economic power is Australia, with imports and exports at about par—$50 billion dollars each. Other countries in the region exhibit comparatively lower trading potential.

Environmental Conditions. The region is faced with problems in virtually all environmental segments, especially marine and coastal degradation as a result of inadequate control of pollutants and ineffective management of its natural resources. Being mostly insular, the countries in the region are particularly sensitive to marine pollution.

In Australia soil erosion results from overgrazing, industrial development, urbanization, and desertification. Cyclones are experienced along the coast. New Zealand has many of the same problems. Earthquakes are common in the region, but are not devastating. Many countries in the region are parties to the Antarctic Environmental Protocol and to treaties governing biodiversity, endangered species, hazardous waste, ozonelayer protection, and similar concerns.

8.2.4 North, South, and Central America

The regions of the Americas vary from one location to another; Table 8.6 presents fair summaries of their environmental conditions. Specific regional issues are elaborated further in the following sections.

South America. The region comprises Argentina, Bolivia, Brazil, Chile, Columbia, Ecuador, French Guiana, Paraguay, Peru, Suriname, Trinidad and Tobago, Uraguay, and Venezuela. The population is approximately 312 million,

Table 8.6 Environmental Condition Ratings in North, South and Central America

Region	Rating	Score
South America	This region has vast resources, yet they are in danger. It has some good initatives, but it remains unclear whether these will be carried out, or whether they will be just for show. This area has perhaps the greatest opportunity for real improvement.	3, 4
Central America	There are good polices in place, but the costs of war are high.	1, 2
North America	This area has some of the strongest environmental regulations in the world. North Americans may be on their way to more environmental awareness.	1

Key to Green Statistics Environmental Issue

Conservation issues have become major popular and political concerns. National and international conservation organizations have gained support from governments, international development agencies, corporations, and commercial organizations in the creation of international conservation treaties. *Earth Journal* uses membership in these environmental conventions as a measure of a region's environmental commitment.

Source: Earth Journal, 1993, Environmental Almanac and Resource Directory, from the Editors of Buzzworm Magazine.

with a total surface area of approximately 6.9 million square miles. The region is rich in natural resources and has an abundance of rain forests, of which the Amazonian rain forest is typical; it also has immense stretches of arid coastal lands with sandy beaches. There is diversity in the topography, climate, and geological features. Overall, the region's climate is tropical and humid. The environmentally protected areas include 2 percent in Brazil, 16 percent in Chile, 38 percent in Ecuador, 0 percent in French Guinea, 3 percent in Paraguay, 10 percent in Venezuela, 5 percent in Colombia and Surinane, about 4 percent in Argentina, Bolivia, and Peru, and 1 percent in Guyana and Uruguay. The predominant natural resources are timber, minerals, oil, copper, iron ore, tin, zinc, gold, silver, bauxite, uranium, antimony, bismuth, sulfur, and tungsten.

Political Conditions. Most countries of the Americas were colonized by Spain and Portugal. Their governments are mostly democratic. The church (Roman Catholic) and labor unions play active roles in the political activities of the region.

Economic Indicators. Natural resources abound in this region. Brazil, the largest country, has a GDP of $886 billion (1994 estimate); Argentina, the second largest, $270 billion. Agriculture looms large in the regional economy. Brazil's exports, as of 1994, totaled $43.6 billion; imports, $33 billion; Argentina's, $16 billion in exports, $21 billion in imports.

Business Organizations. The region is fairly opened-minded to foreign trade and international business cooperation.

Trade. Among the region's exports are iron ore, soybean bran, orange juice, footwear, and automobile parts. Some countries produce crude oil; most others import. Other imports are chemical products and coal. Trading partners include the United States, the European Community (EC), Latin American, the Middle East, and Japan.

Environmental Conditions. There are problems of air and water pollution, desertification, flooding, and erosion. Some areas experience earthquakes and violent windstorms. Most countries in the region have signed international treaties protecting the environment.

Central America. This region includes Antigua, Barbuda, the Bahamas, Barbados, Belize, the Commonwealth of Dominica, Costa Rica, Cuba, the Dominican Republic, El Salvador, St. Kitts and Nevis, Guatemala, Haiti, Honduras, Jamaica, Nicaragua, Panama, St. Lucia, St. Vincent, and the Grenadines. The total population of the region is approximately 57 million. The total surface land area is approximately 280,000 square miles. The region consists mainly of islands with tropical forests that amount to about 30 percent of the total land area, and the climate is hot, humid and very rainy. Most of the islands have been formed from volcanic activities, which have created fertile volcanic soil. Natural resources consist of salt, timber, aragonite, crude oil, nickel, gold, silver, lead, zinc, bauxite, gypsum, and copper.

Political Conditions. Although most countries are tranquil, political conditions in some areas have been turbulent. Cuba is the only communist country in the hemisphere. Nicaragua and Haiti have emerged from dictatorship.

Economic Indicators. Tourism is important in almost all Caribbean islands. Jamaica's GDP is about $8 billion; Panama's, $12 billion; and El Salvador's, $10 billion.

Business Organizations. The islands are very proactive in foreign investments and are, therefore, fairly open to business activities. Foreign investors and consultants are not restricted in many business activities. Group economic activities are strongly influenced by the Caribbean Community and Common Market (CARICOM). Formed in 1973, this organization replaced the Caribbean Free Trade Association (CARIFTA), which was founded in 1965.

Trade. Export products include bananas, coffee, and sugar cane. Imports are, in part, raw materials and consumer and capital goods.

Environmental Conditions. The region's environmental condition has been adversely affected by inadequate industrial pollution control and poor environmental management. Countries in the region are signatories to most international environmental treaties. Poverty and population increases are contributing to environmental degradation in the region.

APPENDICES

APPENDIX I

ORGANIZATIONS AND FUNCTIONAL UNITS OF THE UNITED NATIONS

The United Nations consists of major distinct functional units and departments and/or agencies that have specific and specialized tasks. The specialized task departments and agencies undertake environmental activities such as natural resources management and programs that address biodiversity, climate change, and protection of human health and the environment.

FUNCTIONAL UNITS OF UNITED NATIONS

- UN Secretariat
- Economic and Social Council
- UN Center for Human Settlements
- UN Children's Fund
- UN Conference on Environment and Development (UNCED)
- UN Development Program (UNDP)
- UN Environment Program (UNEP)
- UN Population Fund
- UN Institute for Training and Research
- UN Research Institute for Social Development
- UN University
- World Food Council
- World Food Program

SPECIALIZED TASK DEPARTMENTS

Food and Agriculture Organization of the UN General Agreement on Tariffs and Trade, International Atomic Energy Agency

International Civil Aviation Organization

International Fund for Agricultural Development

International Labor Organization

International Maritime Organization

UN Educational, Scientific, and Cultural Organization, UN Industrial Development Organization

Universal Postal Union

World Bank Group

World Health Organization

World Meteorological Organization

United Nations Secretariat

The UN Secretariat consists of the following functional units and agencies that are involved in environmental activities and programs:

Administrative Committee on Coordination (ACC)
Room S-3720
United Nations
New York, NY 10017
Phone: (212) 963-1234

African Institute for Economic Development and Planning
B.P. 3186
Dakar, Senegal
Operational centers in Cameroon, Morocco, Niger, Rwanda, and Zambia and a liaison office in New York

Center for Science and Technology for Development (CSTD)
One United Nations Plaza
New York, NY 10017
Phone: (212) 963-860
Tx: 422 311
Fax: (212) 963-4116

Center for Social Development and Humanitarian Affairs (CSDHA)
United Nations Office at Vienna
P.O. Box 500
A-1400 Vienna, Austria
Phone: (43 1) 211-310
Tx: 135612
C: UNATIONS

Department of International Economic and Social Affairs (DIESA)
United Nations
New York, NY 10017

Department of Technical Cooperation for Development
United Nations
New York, NY 10017

Economic and Social Commission for Asia and the Pacific (ESCAP)
United Nations Building
Rajdamnern Avenue
Bangkok 10200, Thailand
Phone: (66 2) 2829161-200
Tx: 82392 escap th.
C: ESCAP
Fax: (66 2) 2829602

Economic and Social Commission for Western Asia (ESCWA)
P.O. Box 927115
Amman, Jordan. Phone: (962 6) 694-351
Tx: 21691 unecwa.
Fax: (9626) 694-981
P.O. Box 5747
New York, NY 10163

Economic and Social Council (ECOSOC) Commission on Human Rights
Includes the United Nations Working Group on Indigenous Peoples
Palais des Nations
CH-1211 Geneva 10, Switzerland. Phone: (41 22) 734-6011

Economic Commission for Africa (ECA)
P.O. Box 3001
Addis Ababa, Ethiopia
Phone: (251 1) 5172-00
Tx: 21029, C: ECA
Established 1958. Maintains a Division of Natural Resources

Economic Commission for Europe (ECE)
Palais des Nations
CH-1211 Geneva 10, Switzerland
Phone: (41 22) 734-601 1
Tx: 41 29 62 UNO CH
C:UNATIONS
Fax: (41 22) 734-9825

Economic Commission for Latin America and the Caribbean (ECLAC)
(Commission Economica para America Latina y el Caribe) (CEPAL)
Casilla 179-D
Santiago, Chile
Phone: (56 2) 2085051
Tx: 2340295

C: UNATIONS
Fax: (56 2) 2081946

Office for Ocean Affairs and the Law of the Sea
Office of the Secretary General
United Nations
New York, NY 10017

Office of the United Nations Disaster Relief Coordinator (UNDRO)
Palais des Nations
CH-1211 Geneva 10, Switzerland
Phone: (41 22) 734-601 1
Tx: 41422 dro ch.
C: UNDRO
Fax: (41 22) 733-5623

United Nations Center for Human Settlements
(Habitat) (UNCHS)
P.O. Box 30030
Nairobi, Kenya
Phone: (254 2) 333930
Tx: 22996
C: UNHABITAT
Fax: (2542) 521160

United Nations Children's Fund (UNICEF)
3 United Nations Plaza
New York, NY 10017
Phone: (212) 326-7000
Tx: RCA 239521
C: UNICEF
Fax: (212) 888-7465

United Nations Conference on Environment and Development (UNCED)
Secretariat, B.P. 80
CH-1231 Conches, Switzerland
Phone: (41 22) 789-1676
Fax: (41 22) 7893536

United Nations Conference on Trade and Development (UNCTAD)
Palais des Nations
CH-1211 Geneva, Switzerland
Phone: (41 22) 734-6011
Tx: 289696
C: UNATIONS
Fax: (41 22) 733-6542

United Nations Development Programme (UNDP)
1 United Nations Plaza
New York, NY 10017
Phone: (212) 906-5000
Tx: 125 980
C: UNDEVPRO
Fax: (212) 826-2057

United Nations Nongovernmental Liaison Service (NGLS)
United Nations, New York, NY 10017
Phone: (212) 963-3125
Fax: (212) 963-8712
Palais des Nations
CH-1211 Geneva 10, Switzerland

United Nations Environmental Program (UNEP)

UNEP's jurisdiction includes 9 programs.

- Environment and Development Program
- Environmental Awareness Program
- Earthwatch Program
- Oceans Program
- Water Program
- Terrestrial Ecosystems Program
- Arid and Semi-Arid Lands Ecosystems Program
- Health and Human Settlements Program
- Arms Race and the Environment Program

UNEP Offices and Addresses

Arab League Liaison Office
P.O. Box 130
1012 Tunis, Tunisia
Phone: (216 1) 281-144
Tx: 4846 unep tn
Fax: (216 1) 781-801

Coordinating Unit for the Mediterranean Action Plan
Leoforos Vassileos Konstantinou 48
Athens 501/1, Greece
T: (30 1) 72 44 536

Tx: 222 61 i medugr
C: UNITERRA

Information Unit on Climate Change (IUCC)
UNEP, Palais des Nations
CH-1211 Geneva 10
Switzerland. Phone: (41 22) 758-2527
Fax: (41 22) 758-1189

Intergovernmental Panel on Climate Change (WMO-UNEP)
World Meteorological Organization
Secretariat of the United Nations Scientific Committee on the Effects of Atomic Radiation (UNSCEAR)
Vienna International Centre
P.O. Box 500
A-1400 Vienna, Austria

International Register of Potentially Toxic Chemicals (IRPTC)
Palais des Nations
CH-1211 Geneva, Switzerland
Phone: (41 22) 798-5850
Tx: 28877
Fax: (41 22) 733-2673

Regional Coordination Unit (Caribbean)
14-21 Port Royal Street
Kingston, Jamaica
Phone: (1 809) 922-9269
Tx: 2340 unlos ja

Regional Office for Asia and the Pacific, ESCAP
Rajadamnern Avenue
Bangkok 10200, Thailand
Phone: (66 2) 829161-200
Tx: 82392 th
C: UNITERRA

Regional Office for Europe
Pavillons du Petit Saconnex, 16
Avenue Jean Trembley
CH- 1209 Geneva, Switzerland
Phone: (41 22) 798 84 00
C: UNITERRA

Regional Office for Latin America
Edificio Naciones Unidas
Boulevard de los Virreyes No. 155
Colonia Lomas Virreyes
11 000 Mexico, D.F.
Mexico
Phone: (52 5) 202-4841
Tx: 017-71-055 eciame
Fax: (52 5) 202-0950

Regional Office for North America
Room DC2-0816, United Nations
New York, NY 10017
Phone: (212) 963-8138
C: UNATIONS
1889 F Street N.W.
Washington, DC 20006
Phone: (202) 289-8456
Tx: 89-506 uninfocen wsh

Regional Office for West Asia
P.O. Box 10880
Manama, Bahrain
Phone: (973) 27 60 72
Tx: 7457 unep bn
C: UNEPROWA
Fax: (973) 27 60 75

Secretariat for the Basel Convention (UNEP/ISBC)
Case Postale 59
CH-1292 Chambesy-Geneva, Switzerland
Phone: (41 22) 758-2510
Tx: 415465 une ch
Fax: (41 22) 758-1189

Secretariat for the Convention on International Trade in Endangered Species of Wild Fauna and Flora (UNEP/CITIES)
T: (41 21) 20 00 81
Tx: 24584 ctes ch
C: CITES
Fax: (41 21) 20 00 84

Secretariat of the Convention on the Conservation of Migratory Species of Wild Animals (UNEP/CMS)
Wissënschaftszëntrum
Ahrstrasse 45
D-5300 Bonn 2, Germany

Phone: (49 228) 302 152
Tx: 855-420 wz d
Fax: (49 228) 302 270

UNEP Industry and Environment Office
tour Mirabeau, 39-43 Quai Andr6 Citro@n, F-75739 Paris C6dex 15, France
Phone: (33 1) 45 78 33 33
Tx: 650273
Fax: (33 1) 45 78 32 34

UNITAR/UNDP Center on Small Energy Resources
Via del Corso 303
1-00186 Rome, Italy. T: (39 6) 841-528
C: foodagri i unitar
Established 1984. Cosponsored by UNITAR and the UN Development Program. Promotes the development of small energy resources, particularly for the benefit of rural populations of developing countries, through seminars and exchange of technical information. Branch office in Lima, Peru.

United Nations Environment Program (UNEP)
P.O. Box 30552
Nairobi, Kenya
Phone: (254 2) 333930 or 52000
Tx: 22068 or 22173
C: UNITERRA
Fax: (254 2) 52071 1
(Regional office for Africa, at the Nairobi headquarters).

United Nations Institute for Training and Research (UNITAR)
801 United Nations Plaza
New York, NY 10017
Phone: (212) 963-8621
Tx: 220379 unitr ur
C: UNINSTAR
Fax: (212) 697-8660

United Nations Population Fund (UNFPA)
220 E. 42nd Street
New York, NY 10017
Phone: (212) 297-5000
Tx: 422031 unfpa
C: UNFPA. Fax: (I 212) 370-0201
Established 1969.

United Nations Research Institute for Social Development (UNRISD)
Palais des Nations
CH-1211 Geneva 10, Switzerland
Phone: (41 22) 798-8400
Tx: 28 96 96
C:UNATIONS
Fax: (41 22) 740-0791

United Nations University (UNU)
Toho Seimei Building
15-1 Shibuya 2-chome, Shibuya-ku
Tokyo 150, Japan
Phone: (81 3) 3499281 1
Tx: J25442
C: UNATUNIV
Fax: (81 3) 3499-2828
Established 1973.

UNRISD Programme on Strategies for the Future of Africa
B.P. 3501
Dakar, Senegal
Phone: (221) 21 11 44
Publishes *Research Notes* (bulletin), research and program reports.

UNU Institute for National Resources in Africa (UNU/INRA)
Interim Office
P.O. Box 30592
Nairobi, Kenya
Phone: (254 2) 333930
Tx: 22275 unesco
Fax: (254 2) 520043

UNU Program for Biotechnology in Latin America and the Caribbean (UNU/BIOLAC)
Avenida Principal
Urbanizacion Cumbres de Curumo
Caracas 1080, Venezuela
Phone: (58 2) 772646
Tx: 21279 biunu vc

Fax: (58 2) 9762485

UNU World Institute for Development Economics Research (UNU/WIDER)
Annankatu 42C
SF-00100 Helsinki, Finland
Phone: (358 0) 693841
Tx: 123455 unuei sf
Fax: (358 0) 6938548

World Food Council (WFC)
Via della Terme di Caracalia
I-00100 Rome Italy
Phone: (39 6) -5797 1
Tx: 610181 fao i
C: FOODAGRI
Fax: (39 6) 5745091

World Food Program (WFP)
Via Cristoforo Colombo 426
1-00145 Rome, Italy
Phone: (39 6) 57971
Tx: 6266756
C: WORLDFOOD
Fax: (39 6) 5133537

SPECIALIZED FUNCTIONAL UNITS, AGENCIES, AND DEPARTMENTS OF THE UNITED NATIONS

Food and Agricultural Organization of the United Nations (FAO)
Via Terme di Caracalla
1-00100 Rome, Italy
Phone: (39 6) 57971
Tx: 610181
C: FOODAGRI
Fax: (39 6) 5797-3152

FAO Regional Office for Africa
P.O. Box 1628
Accra, Ghana
Phone: (233) 666851
Tx: 2139
C: FOODAGRI

FAO Regional Office for Asia and the Pacific
Maliwan Mansion
Phra Atit Road
Bangkok 10200, Thailand
Phone: (66 2) 281-7844
Tx: 82815
C: FOODAGRI

FAO Regional Office for Europe
At FAO Headquarters

FAO Regional Office for Latin America and the Caribbean
Casilla 10095
Santiago, Chile
Phone: (56 2) 228-8056
Tx: 340279
C: FOODAGRI

FAO Regional Office for the Near East
P.O. Box 100, Dokki
Cairo, Egypt
Phone: (20 2) 702-229
Tx: 21055
C: FOODAGRI

FAO Liaison Office for North America
1001 22nd Street
N.W., Suite 300
Washington, DC 20437
Phone: (202) 653-2400
Tx: 64255
C: FOODAGRI

FAO Liaison Office with United Nations
Headquarters, Room DCI-1125
United Nations
New York, NY 10017
Phone: (212) 963-6036
Tx: 236350
C: FOODAGRI
Fax: (I 212) 888-6188

Office of the FAO Representative to the United Nations Organizations in Geneva
Palais des Nations
CH-1211 Geneva 10, Switzerland
Phone: (41 22) 734 60 1 1
Tx: 289375
C: FOODAGRI

General Agreement on Tariffs and Trade (GATT)

154 rue de Lausanne
CH-1211 Geneva 21, Switzerland
Phone: (41 22) 739-511 1
Tx: 28787
C: GATT
Fax: (41 22) 731-4206

International Atomic Energy Agency (IAEA)
P.O. Box 100
A-1400 Vienna, Austria
Phone: (43 1) 2360-0
Tx: 1-12645 atom a
C: INATOM
Fax: (43 1) 234564

International Laboratory of Marine Radioactivity
2 Avenue Pri.nce Hérédítaire Albert
MG98000, Monaco
Phone: (33 93) 25 7347
Tx: 479378 ilmr
Fax: (33 93) 25 73 46

International Civil Aviation Organization (ICAO)
1000 Sherbrooke Street West
Montreal, Quebec H3A 2R2, Canada
Phone: (I 514) 285-8219
Tx: 0524513
C: ICAO
Fax: (I 514).288-4772
ICAO offices in Paris, France; Bangkok, Thailand; Cairo, Egypt; Mexico City; Lima, Peru; Nairobi, Kenya; and Dakar, Senegal. Publishes technical documents; provides training.

International Fund for Agricultural Development (IFAD)
Via del Serafico 107
I-00142 Rome, Italy
Phone: (39 6) 54591
Tx: 620330
C: IFAD
Fax: (396)5043463

International Labor Organization (ILO)
4 Route des Morillons
CH-1211 Geneva 22, Switzerland
Phone: (41 22) 799-61 1 1
Tx: 415 647
C: INTERLAB
Fax: (41 22) 798-8685

ILO Liaison Office with the United Nations
220 E. 42nd Street, Suite 3101
New York, NY 10017
Phone: (212) 697-0150
C: LABORINTER
Fax: (212) 883-0844

ILO Regional Office for Africa
P.O. Box 2788
Addis Ababa, Ethiopia

ILO Regional Office for Latin America and the Caribbean
Apartado Postal 3638
Lima 1, Peru
Phone: (51 14) 41-98-00

ILO Regional Office for Asia and the Pacific
P.O. Box 1759
Bangkok, Thailand
Phone: (66 2)

International Maritime Organization (IMO)
4 Albert Embankment
London SE1-7SR, England
Phone: (44 71) 735 761 1
Tx: 23588
C: INTERMAR
Fax: (44 71) 587 3210

United Nations Educational, Scientific, and Cultural Organization (UNESCO)
7 Place de Fontenoy
F-75700 Paris, France
T: (33 1) 45 68 10 10
Tx: 270 602
C: UNESCO
Fax: (33 1) 45 67 16 90
Regional Offices for Science and Technology in Nairobi, Kenya; Montevideo, Uruguay; Amman, Jordan; New Delhi, India; Jakarta, Indonesia; Beijing, China;

and Regional Offices for Education in Dakar, Senegal; Santiago, Chile; Caracas, Venezuela; Amman, Jordan; and Bucharest, Romania.

United Nations Industrial Development Organization (UNIDO)
P.O. Box 300
Vienna International Centre
A-140 Vienna, Austria
Phone: (43 1) 211310
Tx: 135612
C: UNIDO
Fax: (43 1) 232156

Universal Postal Union (UPU)
Case Postale
CH-3000 Berne 15, Switzerland
Phone: (41 31) 43 22 1 1
Tx: 91276 1
C: UPU
Fax: (41 31) 43 22 10

United Nations Industrial Development Organization (UNIDO)
P.O. Box 300
Vienna International Centre
A-1400 Vienna, Austria
Phone: (43 1) 211310
Tx: 135612
C: UNIDO
Fax: (43 1) 232156

Universal Postal Union (UPU)
Case Postale
CH-3000 Berne 15, Switzerland
Phone: (41 31) 43 22 1 1
Tx: 91276 1
C: UPU
Fax: (41 31) 43 22 10

World Bank Group
1818 H Street N.W.
Washington, DC 20433
Phone: (202) 477-1234
Tx: 64145
C: INTBAFRAD
Fax: (202) 477-6391

World Health Organization (WHO)
20 Avenue Appia
CH-1211 Geneva 27, Switzerland
Phone: (41 22) 791-211 1
Tx: 41 54 16
C: UNISANTE
Fax: (41 22) 791-0746
Regional offices in Brazzaville, Congo; Alexandria, Egypt; Copenhagen, Denmark; New Delhi, India; and Manila, Phillipines, as well as liaison offices in New York and Addis Ababa, Ethiopia. Office of the International Agency for Research on Cancer in Lyon, France.

World Meteorological Organization (WMO)
Case Postale 2300
CH-1211 Geneva 20, Switzerland
Phone: (41 22) 730-81 1 1
Tx: 414199
C: METEOMOND
Fax: (41 22) 734-2326

World Tourism Organization (WTO)
Capitan Haya 42
E-28020 Madrid, Spain
T: (341) 5710628
Tx: 42188 omt e
C: OMTOUR
Fax: (341) 5713733

APPENDIX II

FOREIGN COUNTRIES' GOVERNMENTAL MINISTRIES DIRECTLY RESPONSIBLE FOR ENVIRONMENT AND HEALTH

AFGHANISTAN

Ministry of Agriculture and Land Reform
Jamal Minia, Kabul
Phone: (93) 41151
Includes a Forestry and Range Department and a Department of Environmental Protection.

Ministry of Public Health, Kabul
Phone: (93) 40851

Ministry of Water Resources Development and Irrigation
Darulaman Wat, Kabul
Phone: (93) 40743

ALBANIA

Committee for Environmental Protection and Preservation
Tirana
Phone: (355-42) 283-90
Fax (355-42) 346-32

Ministry of Agriculture and Fisheries
Tirana
Includes the Ministry of Forestry.

Ministry of Culture
Tirana

Ministry of Health, Department of Hygiene and Environmental Protection
Tirana

Hydrometeorology Institute
Tirana
Monitors air and water pollution.

Academy of Science of the People's Republic of Albania
Tirana

Albanian Nature Protection Association
University of Tirana
Faculty of Natural Science
Tirana

Phone: (355-42) 248-05
Fax: (355-42) 247-72

ALGERIA

National Agency for the Protection of the Environment
B.P. 154, El-Annaser
Phone: (213) 771414

Ministere de l'Agriculture, Agénce National pour la Protéction de la Nature
(Ministry of Agriculture, National Agency for Protection of Nature)
Jardin Botanique
El Hamm, Algeria

Ministry of Health
25 Boulevard Laola 'Abd al-Rahmane
El-Madania, Algeria
Phone: (213) 663315

Ministry of Water Resources, Forestry, and Fishing
Le Grand Seminaire
Kouba, Algeria
Phone: (213) 589500

Institut des Sciences de la Mer et de l'Amenagement du Littoral
(Institute of Marine Sciences and Coastal Zone Management)
B.P. 90, Algeria
Phone: (213) 648457
Tx: 67523

National Institute of Agronomy
Department of Forestry and Nature Protection
Avenue Pasteur, 116-200 El Harrach-Algeria
Phone: (213) 2761987
Fax: (213) 2731823

National Institute of Forestry Research
B.P. 37, Cheraga-Alger
Phone: (213) 849790

AMERICAN SAMOA

American Samoa Environmental Protection Agency
Office of the Governor
Pago Pago 96799
Phone: (684) 633-2304

National Park Service, Pacific Area Office
Box 50165, 300 Ala Moana Boulevard, Room 6305
Honolulu, Hawaii 96850
Administers American Samoa National Park.

American Samoa Department of Parks and Recreation
U.S. Department of the Interior
P.O. Box 3809, Pago Pago 96799
Responsibilities include administration of natural preserves and territorial parks.

ANDORRA

Council for Agriculture and Natural Heritage
Larrer Maragall S/N
Andorra la Vella

Ministry of Health, Labor and Welfare
C. Prat de la Creu, 62-64
Andora la Vella

Association for the Defense of Nature
Apartat de Correus Espanyols, no. 96
Andorra la Vells

ANGOLA

Ministry of Agriculture and Rural Development
Avenida Norton de Matos 2
Luanda
Phone: (244) 32 39 34
Tx: 3322 minagri an

Ministry of Health
Rua Diego Cao
Luanda

Ministry of Industry
Predio Gomes Irmaos
Luanda
Includes the National Department of Water.

Ministry of Fisheries

Avenida 4 de Fevereiro
Predio Atlantico

Angolan Environmental Association
c/o Faculdade de Ciencias (Biologica)
Universidade Agostinho Neto
C.P. 815, Luanda
Phone: (244) 345000
Tx: 3207 game k an
Fax: (244) 345000, ext. 221

ANGUILLA

British Territory

Ministry of Tourism and Natural Resources
Department of Agriculture and Fisheries
The Secretariat
The Valley, Anguilla
Phone: (1-809) 497-2451
Tx: 9313

ANTIGUA AND BARBUDA

Ministry of Agriculture, Lands, and Fisheries
High Street
St. John's, Antigua
Phone: (1-809) 462-1007

National Parks Authority
P.O. Box 1283
St. John's, Antigua
Phone: (1-809) 460-1379
Fax: (1-809) 460-1516

Historical, Conservation, and Environmental Commission (HCEC)
c/o Ministry of Economic Development, Tourism, and Energy
Queen Elizabeth Highway
St. John's, Antigua
Phone: (1-809) 462-0092
Fax: (1-809) 462-2836

Environmental Awareness Group (EAG)
c/o The Museum of Antigua and Barbuda
St. John's, Antigua
Phone: (1-809) 463-1060
Fax: (1-809) 463-1532

Historical and Archaeological Society of Antigua and Barbuda
P.O. Box 103
St. John's, Antigua
Phone: (1-809) 462-3946

ARGENTINA

State Secretariat of Natural Resources and Human Environment
San Martin 459
1004 Buenos Aires
Phone: (54 1) 394-6612

Ministry of Public Health and Social Order
Defensa 120
1345 Buenos Aires
Phone: (1) 304-322

Argentine Institute of Arid Zone Research
Casilla de Correo 507
5500 Mendoza
Phone: (54 61) 24-1995
Tx: 55438 cytme ar

National Council of Scientific and Technical Research
Rivadavia 1917
1033 Buenos Aires
Phone: (54 1) 953-7230
Tx: 18052 cicyt ar

Center for Studies and Planning of the Environment
Calle 53, No. 506, La Plata
1900 Buenos Aires
Phone: (54 1) 32-601

ARMENIA

Ministry of Nature and Environmental Protection
35 Moscoviana Street
Yerevan
Phone: (8852) 53-0741

Ministry of Material Resources
3 Khorentsi Street
Yerevan
Phone: (7 873) 538082

Ministry of Energy and Fuel

Republic Square, Government Building 2
Yerevan
Phone: (7 873) 521964

ARUBA

Ministry of Public Works,
Paarden
Oranjestad
Phone: (273) 271080

Ministry of Health
Hospitaalstraat 4
Oranjestad
Phone: (297 8) 24200

Ministry of Energy Affairs
Oranjestad

AUSTRALIA

Ministry of Health
Suite MG60, Parliament House
Canberra, ACT2600
Phone: (6) 273-7220
Fax: (6) 273-4146

Department of Industrial Relations
Suite MF27, Parliament House
Canberra, ACT 2600
Phone: (6) 273-7320
Fax: (6) 273-4115

AUSTRIA

Ministry for Environment, Youth, and Family
Federal Environmental Agency
Radetzkystrasse 2
A-1030 Vienna
Phone: (1) 711580
Fax: (1) 711584221

Ministry of Agriculture and Forestry
Stubenring 1
A-1012 Vienna
Phone: (1) 711000
Fax: (1) 71139311

Ministry of Foreign Affairs
Ballhausplatz 2
A-1014 Vienna
Phone: (1) 531150
Fax: (1) 5354530

AZERBAIJAN

Office of the President
Lermontova 63
Baku
Phone: (8-922) 92-77-38

Azerbaijan Green Movement
4, 370005 Baku
Phone: (1) 39-91-95
Tx: 142178 bakin su
Fax: (1) 98-73-03

BAHAMAS

Ministry of Agriculture, Trade, and Industry
P.O. Box N-3028
Nassau
Phone: (I 809) 325-7502

Ministry of Health, Department of Environmental Health Services
P.O. Box N73729
Nassau
Phone: (1-809) 322-4908
Tx: 20264

Ministry of Works and Lands, Department of Lands and Surveys
P.O. Box N-592
Nassau
Phone: (1-809) 322-2328

Bahamas National Trust
P.O. Box N-4105
Nassau
Phone: (1-809) 393-1317
Fax: (1-809) 393-4978

BAHRAIN

Environmental Protection Council
P.O. Box 26909
Adliya
Phone: (973) 27 57 92
Tx: 8511 health
Fax: (973) 29 36 94

Environmental Protection Technical Secretariat
P.O. Box 26909
Adliya
Phone: (973) 29 36 93
Tx: 8511 health bn
Fax: (973) 29 36 34

Ministry of Commerce and Agriculture
P.O. Box 5479
Manama
Phone: (973) 53 15 31
Tx: 9171

Ministry of Health
P.O. Box 12
Manama
Phone: (973) 25 26 05
Tx: 8511 health bn
Fax: (973) 25 25 69

Arabian Gulf University
Desert and Arid Zones Sciences Program
Manama
Phone: (973) 27 72 09
Tx: 7319

BANGLADESH

Ministry of Environment and Forest, Livestock and Fisheries
Dhaka 12

Ministry of Irrigation, Water Development, and Flood Control
Bhaban No. 6, Secretariat, Dhaka
Phone: (880 2) 414440

Bangladesh Academy of Sciences
3/8 Asad Avenue
Dhaka 7
Phone: (8-802) 310425
C: SCIENCE

BARBADOS

Ministry of Labor, Consumer Affairs and the Environment
Marine House
Hastings, Christ Church
Phone: (1-809) 427-5420

Ministry of Community Development and Culture
Department of Community Development
Marine House
Hastings, Christ Church
Phone: (1-809) 4275420

Ministry of Agriculture, Food, and Fisheries
Graeme Hall
Christ Church
Phone: (1-809) 4284150

Ministry of Housing and Lands
Marine House
Hastings, Christ Church
Phone: (1-809) 427-5420
Tx: 386222

Ministry of Health
Jemmotts Lane
St. Michael
Phone: ((1-809) 426-5080
Fax: (1-809) 426-5570

Ministry of Public Works, Communication and Transportation
P.O. Box 25
Bridgetown
Phone: (1-809) 429-2191

Barbados Environmental Association
P.O. Box 132
Bridgetown
Phone: (1-809) 427-0619
Fax: (1-809) 427-0619

Barbados National Trust
Ronald Tree House
2 Tenth Avenue
Belleville, St. Michael
Phone: (1-809) 426-2421

University of the West Indies
Center for Resource Management and Environmental Studies
P.O. Box 64
Bridgetown
Phone: (1-809) 425-1310

BELARUS

Ministry of Resources
Kazintsa
220855 Minsk
Phone: (0172) 26-00-84

Ministry of Water Resources and Land Reclamation
Minsk
Phone: (0172) 34-59-93

State Committee on Ecology
Minsk
Phone: (7 0172) 20 66 91

Belarussian Ecological Union
Minsk
Phone: (7 0172) 278791
265942

BELGIUM

Walloon Region
Department of Environment, Agriculture, and Housing
87 Avenue Albert ler
B-5000 Namur
Phone: (32 81) 24 66 1 1
Fax: (32 8 1) 23 14 68

Flemish Region
Department of the Environment, Nature Protection and Rural Planning
Kunstlaan 43
B-1040 Brussels
Phone: (32-2) 513 74 64
Fax: (32-2) 514 27 01

Brussels Region
Administration of Natural Resources and the Environment
Trierstraat 49
B-1040 Brussels
Phone: (32 2) 231 12 55
Fax: (32 2) 230 88 15

Ministry of Public Health and the Environment
Rue de la Loi 66
1040 Brussels
Phone: (2) 238-28-11
Fax: (2) 230-38-62

BELIZE

Ministry of Agriculture and Fisheries
Belmopan
Phone: (8) 22131

BELMOPAN

Ministry of Health
Belmopan
Phone: (8) 22362

Ministry of Natural Resources
Belmopan
Phone: (8) 22321

Ministry of Tourism and the Environment
Belmopan
Phone: (8) 23393
Fax: (8) 22862

BENIN

Directorate of Water, Forests, and Hunting
B.P. 393, Cotonou
Phone: (229) 33 06 62
Fax: (229) 33 19 56

Agriculture Directorate
B.P. 58, Porto-Novo
Phone: (229) 21 32 90

General Directorate of Planning
B.P. 239, Cotonou

Ministry of Public Health
B.P. 882, Cotonou
Phone: (229) 31 26 70

Ministry of Environment, Housing and Urban Development
B.P. 01-3621, Cotonou
Phone: (229) 31-46-61

BERMUDA

Ministry of the Environment
Government Administration Building, 3rd Floor
30 Parliament Street
Hamilton HM 12
Phone: (1-809) 295-5151

BHUTAN

Ministry of Agriculture and Forests
P.O. Box 130, Thimphu
Phone: (975) 22503 or 22368
Fax: (975) 22395

National Environmental Secretariat
Planning Commission
Royal Government of Bhutan
Tashichodzong, Thimphu
Phone: (975) 22056
C: BHUTANPLAN
Fax: (975) 23069

BOLIVA

Ministry of Sustainable Development

La Paz
Phone: (2) 372063

National Environment Secretariat
La Paz
Phone: (2) 369304
Fax: (2) 357535

Ministry of Rural and Farming Affairs
La Paz
Phone: (2) 374260

BOSNIA AND HERZEGOVINA

Ministry of Environmental Protection, Planning and Construction
Vojvode Putnika 3
71000 Sarajevo
Phone: (71) 213777
Fax: (71) 653592

Office for the Protection of Cultural Monuments
Department of Nature Conservation
Phone: (71) 653555

BOTSWANA

Ministry of Agriculture
Private Bag 003
Gaborone
Phone: (31) 35081
Tx: 2543

Department of Crop Production, Forestry Division
Phone: (267) 350500
Fax: (31) 356027

Ministry of Commerce and Industry
Department of Wildlife and National Parks
P.O. Box 131
Gaborone
Phone: (31) 350500

Ministry of Local Government and Lands
Private Bag 007
Gaborone
Phone: (31) 3542091

Ministry of Mineral Resources and Water Affairs
Private Bag 0018
Gaborone
Phone: (31) 352454

BRAZIL

Ministry of the Environment and Legal Amazon Region
Palacio do Planalto
Anexo II–Terreo
70150-902 Brasilia–DF
Phone: (61) 211-1230
Fax: (61) 226-9871

Ministry of Science and Technology
Brasilia, DF 70067
Phone: (61) 236-2269

BRUNEI

Ministry of Primary Resources and Industry
Includes:
Department of Forestry
Bandar Seri Begawan 2067
Phone: (2) 244545
Fax: (2) 242727

Department of Fisheries
Bandar Seri Begawan 2069
Phone: (2) 242067
Fax: (2) 242069

Ministry of Health
Bandar Seri Begawan 1210
Phone: (2) 226640
Fax: (2) 240980

Ministry of Development
Town and Country Planning Commission
Bandar Seri Begawan 1190
Phone: (2) 244591
Fax: (2) 242313

BULGARIA

Ministry of the Environment
William Gladstone 67
1000 Sofia
Phone: (2) 87 61 51
Fax: (2) 52 16 34

BURKINA FASO

Ministry of the Environment and Tourism
B.P. 7044, Ouagadougou
Phone: 33 32 13 or 30 63 94

Ministry of Health
B.P. 7009, Ouagadougou
Phone: 30 69 95 or 30 70 77

BURUNDI

Ministry of Planning, Tourism, and Environment
B.P. 1830, Bujumbura

National Institute for the Environment and Conservation of Nature
B.P. 56, Bujumbura
Phone: (2) 206020

Ministry of Agriculture and Livestock
Department of Water and Forests
B.P. 631, Bujumbura

Ministry of Public Health
Bujumbura
Phone: (2) 26020

CAMBODIA

Ministry of Agriculture, Forests and Fisheries
Phnom Penh
Phone: (23) 23689

CAMEROON

Ministry of Planning and Regional Development
Department of Regional Development and Environment
Yaounde
Phone: (23) 23 40 40

Ministry of Agriculture and Forestry Department
B.P. 194, Yaounde
Phone: (23) 23 40 85

Ministry of Tourism, Wildlife and Protected Areas Department
B.P. 266, Yaounde
Phone: (23) 22 44 11

Ministry of Higher Education, Computer Services, and Scientific Research
Center of Forestry Research
B.P. 2102, Yaounde

Ministry of Public Health
Yaounde
Phone: (23) 22 20 01

Ministry of Livestock, Fisheries, and Animal Husbandry
Yaounde
Phone: (23) 22 33 11

Ministry of Mines, Water, and Energy
Yaounde
Phone: (23) 23 34 04

CANADA

Environment Canada (Department of the Environment)
Ottawa, Ontario
Phone: (1-819) 997-2800
Fax: (1-819) 953-6789

Forestry Canada (Department of Forestry/Science and Sustainable Development Directorate)
Ottawa, Ontario KIA OC5
Phone: (1-819) 997-1107

Agriculture Canada (Department of Agriculture)
Ottawa, Ontario KIA OC7
Phone: (1-613) 995-8963
Fax: (1-613) 996-9564

Department of Energy, Mines, and Resources
G 580 Booth Street
Ottawa, Ontario KIA OE4
Phone: (1-613) 995-0947

Department of Fisheries and Oceans
200 Kent Street
Ottawa, Ontario KIA OE6
Phone: (1-613) 993-0600

Department of Health and Welfare
Ottawa, Ontario KIA OK9
Phone: (1-613) 957-2991
(Environmental health.)

Indian and Northern Affairs Canada (Department of Indian and Northern Affairs)
Ottawa, Ontario KIA OH4
Phone: (1-613) 997-9885
Fax: (1-613) 997-1587

Atomic Energy Control Board
P.O. Box 1046

Ottawa, Ontario KIP 5S9
Phone: (1-613) 995-6941

Canadian International Development Agency (CIDA)
200 Promenade du Portage
Hull, Quebec KIA OG4
Phone: (1-819) 997-6100

International Development Research Centre (IDRC)
P.O. Box 8500
Ottawa, Ontario KIG 3H9
Phone: (1-613) 236-6163

CAPE VERDE

Ministry of Rural Development and Fisheries
C.P. 50, Praia
Phone: (238) 335

National Institute of Agrarian Research
C.P. 84, Praia

Ministry of Health and Social Affairs
C.P. 47, Praia
Phone: 422

CAYMAN ISLANDS

Natural Resources Laboratory
Cayman Islands Government
P.O. Box 486
Grand Cayman

CENTRAL AFRICAN REPUBLIC

Ministry of Water, Forests, Hunting, Fisheries, and Tourism
B.P. 830, Bangui
Phone: (236) 61 24 31

Ministry of Health and Social Affairs
Bangui

Ministry of Public Works and Town Planning
Bangui
Phone: (236) 61 28 00

State Secretariat for Scientific and Technical Research
B.P. 731, Bangui
Phone: (236) 61 21 07

CHAD

Ministry of Tourism and the Environment
B.P. 447, N'Djamena
Phone: (235) 51 56 56

Ministry of Agriculture and the Environment
B.P. 436, N'Djamena
Phone: (235) 51 37 52

Division of National Parks and Wildlife Reserves
B.P. 905, N'Djamena
Phone: (235) 51 23 05
Fax: (235) 51 43 97

CHILE

Ministry of Agriculture
Teatinos 40
Piso Santiago
Phone: (56 2) 696-3241

National Corporation for Forestry and Protection of Renewable Natural Resources
Santiago
Phone: (56 2) 696-0783

Ministry of Health, Department of Environmental Programs
Avenue Miguel Claro 1194
Santiago
Phone: (56 2) 381-843

Chilean Antarctic Institute
Santiago
Phone: (56 2) 231-0105

Institute of Fisheries Development
Santiago
Phone: (56 2) 225-2331

CHINA

Ministry of Urban and Rural Construction and Environmental Protection
National Environmental Protection Agency (NEPA)
Phone: (86 1) 653681
Fax: (86 1) 601-5641

National Information Center of Environmental Science and Technology (NICEST)

Bai Wan Zhuang
Beijing 100835
Phone: (1) 839-3934
Fax: (1) 839-3245

Ministry of Agriculture
Hepingli, Dongcheng Qu
Beijing
Phone: (1) 463061
Fax: (1) 5002448

Ministry of Forestry
Hepingli, Dongcheng Qu
Beijing
Phone: (86 1) 463061

Ministry of Public Health
44 Houbaibeiyan
Beijing
Phone: (86 1) 440531
Tx: 22193

Ministry of Water Resources and Electric Power
1 Xiang Lane, Baiguang Je
Guanganmen, Beijing
Phone: (86 1) 365563

Ministry of Geology and Mines
Xisi Yangshi Jie
Beijing
T: (86 1) 668741
Tx: 22531

COLOMBIA

National Department of Planning
Calle 25, No. 13-19
Bogota
Phone: (57 1) 282-4055

Ministry of Agriculture
Carrera 10A, No. 20-10, Piso
Santa Fe de Bogata, D.C.
Phone: (57 1) 241-9095

National Institute of Renewable Natural Resources and the Environment
Bogota
Phone: (57 1) 285-4417
Fax: (57 1) 283-3458

Ministry of National Education
Department of Environmental Education
Bogota
Phone: (57 1) 222-2800
Fax: (57 1) 222-0324

National Corporation of Forestry Research and Development
Bogota
Phone: (57 1) 267-6844

COMOROS

Ministry of Planning
Environment Directorate
B.P. 12, Moroni

Ministry of Public Health
B.P. 42
Phone: (73) 22-77

CONGO

Ministry of Agriculture, Water, Forests, Environment, and Biodiversity
B.P. 98, Brazzaville
Phone: (242) 83 18 47 or (242) 83 24 58

Ministry of Hydrocarbons
B.P. 2120
Centre Administratif, Quartier Plateau
Phone: (1) 83 44 67 or 81 12 81

Ministry of Industrial Development, Mines and Energy
Avenue Amilcar Cabral
A Cote UCB "B"
Centre Ville, Brazzaville
Phone: (1) 83 18 27

COOK ISLANDS

Ministry of Conservation, Conservation Service
P.O. Box 371
Tupapa, Rarotonga
Phone: (682) 21256

COSTA RICA

Ministry of Natural Resources, Energy, and Mines
Avenida 8-10
Calle 25, Apartado 10104
San Jose 1000
Phone: (1) 57-1417 or 33-3878

Ministry of Agriculture and Livestock
Apartado 10094

1000 San Jose

Ministerio de Salud (Ministry of Health)
Apartado 10123
San Jose
Phone: (506) 23-0333
(Environmental health.)

CÔTE D'IVOIRE

Ministry of Environment, Construction, and City Planning
National Environment Commission
B.P. V-67, Abidjan
Phone: (225) 22 53 54

Ministry of Agriculture and Animal Resources
General Directorate of Water and Forests
B.P. V94, Abidjan
Phone: (225) 22 16 29

CROATIA

Ministry of the Environment, Physical Planning, and Construction
Avenija Vukovar
Zagreb
Phone: (38 41) 432022

CUBA

Ministry of Agriculture
Avenida Independenceia
Havana
Phone: (1) 7011434

National Commission on the Environment
Phone: (53 7) 290501
Fax: (53 7) 625604

Ministry of Agriculture
Phone: (53 7) 701434
Tx: 511966

CYPRUS

Ministry of Agriculture and Natural Resources
Louki Akrita
Ayios, Nicosia
Phone: (357 2) 30 25 86

Ministry of Health
Nicosia
Phone: (2) 302284

Ministry of Interior
Department of Town Planning and Housing
Nicosia
Phone: (2) 402105

CZECHOSLOVAKIA

Federal Committee for Environment
Slezska 9
CS-129 09 Prague 1
Phone: (42 2) 215-11 1 1
Fax: (42 2) 254-964

Ministry of Environment
Vrsovicka 65
CS-100 10 Prague 10
Phone: (42 2) 7121111
Fax: (422) 731357

Czech Institute for Nature Conservation
CS-129 09
Prague 2
Phone: (42 2) 215-2609
Fax: (42 2) 731357

Slovak Commission for Environment
Hlboka 2
CS-812 35
Bratislava
Phone: (42 7) 492-553
Fax: (42 7) 311-384

Central Board for Nature Conservation
CS-823 00
Bratislava
Phone: (42 7) 687-21

DENMARK

Ministry of the Environment
Slotholmsgade 12
1216 Copenhagen K
Phone: (45) 33 92 33 88
Fax: (45) 33 32 22 27

National Agency of Environmental Protection
Strandgade 29
1401 Copenhagen K
Phone: (45) 31 57 83 10
Fax: (45) 31 57 24 49

National Forest and Nature Agency

Slotsmarken 13
2970 Hoersholm
Phone: (45) 45 76 53 76
Fax: (45) 76 54 77

National Agency for Physical Planning
Haraldsgade 53
2100 Copenhagen 0
Phone: (45) 39 27 11 00
Fax: (45) 39 27 12 66

Danish Environmental Research Institute
Thoravej 8, 3rd Floor
2400 Copenhagen NV
Phone: (45) 31 19 77 44
Fax: (45) 38 33 26 44

Ministry of Foreign Affairs
Department of International Development (DANIDA)
Asiatisk Plads 2
DK-1256 Copenhagen K
Phone: (45) 33 92 00 00
Fax: (45) 31 54 15 33

DJIBOUTI

National Committee for the Environment
Technical Secretariat
B.P. 1938, Djibouti
Phone: (253) 35 28 00
Fax: (253) 35 63 22

Ministry of Commerce, Transport, and Tourism
National Office of Tourism and Crafts
Service for Protection of Sites and Environment
B.P. 1938, Djibouti

Ministry of Agriculture and Rural Development
Agriculture and Forest Service
B.P. 224, Djibouti

Ministry of Health
B.P. 296, Djibouti
Phone: (253) 35 14 91 or 35 33 31

DOMINICA

Ministry of Agriculture and Forestry
Government Headquarters
Roseau
Phone: (1-809) 448-2401

Ministry of Health
Government Headquarters, Roseau
Phone: (1-809) 448-2401

DOMINICAN REPUBLIC

Environmental Commission
Dr. Delgado 58, Apartado 1351
Santo Domingo
Phone: (1-809) 689-3808
Tx: 346-0461

Secretariat of Public Health and Social Welfare
Ensanche La Fe
Santo Domingo
Phone: (809) 566-5988

ECUADOR

National Council of Development
Program of Natural Resources and Environment
Phone: (593 2) 564-518

Ecuadorian Institute of Sanitary Works
Department of Environmental Sanitation
Phone: (593 2) 562-810

Ecuadorian Institute of Water Resources
Juan Larrea 534 y Riofrio
Quito
Phone: (593 2) 545-566

Ministry of Agriculture and Livestock
Av. Amazonas y Eloy Alfaro Edf. MAG
Quito
Phone: (593 2) 541955

Forestry Program
Av. Amazonas y Eloy Alfaro Edf. MAG
Quito
Phone: (593 2) 548-924

National Institute for Colonization of the Amazon Region
AV. Amazonas y la Nina
Quito
Phone: (593 2) 548-338

Ministry of Energy and Mines
Sub-Secretariat of Environment
Madinat Nasr, Abbassia
Santa Prisca 223 y Manuel Larrea
Quito
Phone/Fax: (593 2) 570-341

EGYPT

Environmental Affairs Agency
Cairo
Phone: (20 2) 3416546
Fax: (20 2) 3420768

Ministry of Agriculture and Land Reclamation
Phone: (20 2) 702-677

Suez Canal Authority
Irshad Building
Ismailia
Phone: (20 64) 220000
Tx: 63238
Fax: (20 64) 220784

Ministry of Electricity and Energy
Madinat Nasr
Abbassia
Phone: (2) 834461

EL SALVADOR

National Committee of Environmental Protection
c/o Ministerio de Planificacion
San Salvador
Phone: (503) 71-3266

Ministry of Agriculture and Livestock
Center of Renewable Natural Resources
Boulevard de los Hdroes y 21 Calle Poniente
San Salvador
Phone: (503) 23-2598

Ministry of Public Health and Social Assistance
Calle Arce 827
San Salvador
Phone: (503) 21-0966

EQUATORIAL GUINEA

Ministry of Agriculture, Livestock, Fisheries, and Forestry
Malabo
Phone: (240 9) 21-19

Ministry of Public Works and Town Planning
Malabo
Phone: (240 9) 34-60 or 23-03 or 25-05 or 27-39

Ministry of Health
Malabo

ESTONIA

Ministry of Environment
Toompuiestee 24
Talliunn EE0100
Phone: (2) 452-507
Fax: (7 142) 45 33 10

ETHIOPIA

Ministry of Agriculture and the Environment
Natural Resources Conservation and Development Main Department
Addis Ababa

Ethiopian Wildlife Conservation Organization
P.O. Box 386
Addis Ababa
Phone: (251 1) 154436
Fax: (251 1) 518977

State Forest Conservation and Development Department
P.O. Box 386
Addis Ababa
Phone: (251 1) 518040

Environment Health Division
P.O. Box 5504
Addis Ababa
Phone: (251 1) 441944

FALKLAND ISLANDS

Office of the Governor
Government House
Stanley
Phone: (500) 27433
Fax: (500) 27434

FAROE ISLANDS

Government Offices
P.O. Box 64
1 10 Tershavn
Phone: (298) 11080

Tx: 81310
Fax: (298) 14942
Includes: Ministry of Fisheries
Ministry of Agriculture
Ministry of Health and Social Services

FIJI

Ministry of Forestry
P.O. Box 2218
Suva
Phone: (679) 302 740

Ministry of Health
Private Mail Bag
Tamavua, Suva
Phone: (679) 320 976

Ministry of Housing and Urban Development
P.O. Box 2350
Suva
Phone: (679) 211 416

Department of Town and Country Planning
Phone: (679) 211 208
Fax: (679) 303 515

Ministry of Lands and Mineral Resources
P.O. Box 222
Suva
Phone: (679) 211 515

Ministry of Primary Industries and Cooperatives
Private Mail Bag
Suva
Phone: (679) 312 745

FINLAND

Ministry of the Environment
Eteldesplanadi 18A, PL 399
SF-00121 Helsinki
Phone: (358 0) 19911
Fax: (358 0) 1991399

Office of National Parks
National Board of Forestry
Erottajan Katu
PL 233, SF-00121 Helsinki
Phone: (358 0) 9061631

Ministry of Agriculture and Forestry
PL 232, SF-00171
Helsinki 17

Finnish International Development Agency (FINNIDA)

Ministry of Foreign Affairs
Helsinki
Phone: (358 0) 134161

National Board of Health
Office of Environmental Hygiene
Helsinki
Phone: (358 0) 77231

FRANCE

Ministere de I'Environnement (Ministry of the Environment)
41 Avenue Georges Mandel
F75016 Paris
Phone: (33 1) 46 47 31 32 or 47 58 12 12
Tx: 642849
Fax: (33 1) 46 47 38 95

Ministry of Agriculture and Forests
78 Rue de Varenne
F75700 Paris
Phone: (33 1) 49 55 49 55

FRENCH GUIANA

Prefecture, rue Fiedmond
97307 Cayene Cédex
Phone: (594) 30 05 20
Fax: (594) 30 02 77

FRENCH POLYNESIA

Ministry of the Environment
B.P. 4562
Papeete

Ministry of Health
Papeete

Ministry of Agriculture, Traditional Crafts, and the Cultural Heritage
Papeete

FRENCH SOUTHERN AND ANTARCTIC LANDS

Ministry of Overseas Departments and Territories
and Central Administration of the French Southern and Antarctic Lands
27 Rue Oudinot

F75700 Pairs 07 SP
France
Phone: (1 33) 47 83 01 23

GABON

Ministry of Tourism, the Environment and National Parks
B.P. 3241
Libreville
Phone: (241) 73 17 07

Ministry of Water and Forests
B.P. 2275
Libreville

Ministry of Energy and Hydraulic Resources
B.P. 1172
Libreville

Ministry of Scientific Research
B.P. 496
Libreville

GAMBIA

Ministry of Natural Resources and the Environment
5 Marina Parade
Banjul

Department of Wildlife Conservation
Forestry Department and Environment Unit
Phone: (220) 27307

Ministry of Agriculture
Department of Agricultural Services
Banjul

GEORGIA

Office of the Prime Minister
Prospekt Plekhanova 103
380064 Tbilisi

GERMANY

Federal Ministry of the Environment
Abteilung N. Postfach 12 06 29
D-5300 Bonn 1
Phone: (49 228) 3050
Fax: (49 228) 3053225

Ministry of Agriculture and Forests
Postfach 14 02 70
D-5300 Bonn 1
Phone: (49 228) 5291
Fax: (49 228) 5294262

Federal Ministry for Economic Cooperation
Phone: (49 228) 5351
Fax: (49 228) 535202

Federal Institute for Nature Conservation and Landscape Ecology
Phone: (49 228) 84910
Fax: (49 228) 8491200

GHANA

Ministry of Environmental Protection Council
P.O. Box M326
Ministries Post Office
Accra
Phone: (233 21) 66 46 97

Ministry of Lands and Natural Resources
P.O. Box M212
Accra
Phone: (233 21) 66 54 21

Department of Game and Wildlife
P.O. Box M239
Accra
Phone: (233 21) 66 61 29

Forestry Department
P.O. Box 527
Accra
Phone: (223 21) 77 61 45

Ministry of Agriculture
Forestry Commission
P.O. Box M434
Accra
Phone: (233 21) 22 13 15

Ghana Academy of Arts and Sciences
P.O. Box M-32
Accra
Phone: (233 21) 77 76

African-Environmental Research and Consulting Group (AERCG)
Nonprofit Organization, Box 91
Accra
Phone: (233 21) 772980

Ghana Association for the Conservation of Nature

c/o Institute of Renewable Natural Resources
University of Science and Technology
Kumasi

GREECE

Ministry of Environment, Physical Planning and Public Works
17 Amaliados Street
GR 115 23 Athens
Phone: (30 1) 6431461
Fax: (30 1) 6442682

Directorate of the Environment
Poulion 8
GR 115 23 Athens
Phone: (30 1) 6461189
Tx: 222088

Environmental Planning Division
Nature Management Section
Trikalon 36
GR 115 26 Athens

Ministry of Energy and Natural Resources
Department of Water and Natural Resources
1 Zalokosta Street
GR 106 71 Athens
Phone: (30 1) 3609320

Ministry of Agriculture and Forestry
3-5 Ippokratous Street
Athens
Phone: (30 1) 3637659

GREENLAND

Groenlands Hjemmestyre (Greenland Home Rule Government)
P.O. Box 1015
3900 Nuuk
Greenland
Phone: (299) 23000
Tx: 90613
Fax: (299) 25002

Secretariat for Administration and the Environment
Secretariat for Fisheries and Industry
Office in Denmark
Simleboderne 2
DK- 1 122 Copenhagen K
Denmark
Phone: (45 33) 13 42 24
Tx: 15804
Fax: (45 33) 32 20 24

GRENADA

Ministry of Agriculture, Forestry Division
Archibald Avenue
St. George's
Phone: (1-809) 440-2628

Ministry of Health and Housing and the Environment
St. George's
T: (1-809) 440-2649

Ministry of Tourism, National Parks, and Protected Areas
Carenage
St. George's
T: (1-809) 440-0366

GUADELOUPE

Prefecture
Palais d'Orleans, rue Lardenoy
97109 Basse-Terre Cédex
Phone: (590) 811560
Fax: (590) 91832

Ministere des Departments et Territoires d'Outre-Mer (Ministry of Overseas Departments and Territories)
27 Rue Oudinot
F-75700 Paris 07 SP
France
Phone: (133) 47 83 01 23

GUAM

Environmental Protection Agency
IT&E Harmon Plaza
Complex Unit D-104
130 Rojas Street
Harmon, Guam 9691 1
Phone: (1-671) 646-8863
Fax: (1-671) 646-9402

Department of Agriculture
Agana, Guam 96910
Phone: (1-671) 734-3941

Division of Aquatic and Wildlife Resources
P.O. Box 2950
Agana, Guam 96910
Phone: (1-671) 734-3944

Fax: (1-671) 734-6570

Division of Forestry and Soil Resources
P.O. Box 2950
Agana, Guam 96910
Phone: (1-671) 7343948

Marianas Audubon Society
P.O. Box 4425
Agana, Guam 96910

GUATEMALA

National Environment Commission
7a Avenida 4-35, Zona 1
Guatemala
Phone: (502-2) 21816
Fax: (502-2) 535109

Directorate General of Forests
7a 6-80, Zona 13
Guatemala
Phone: (502 2) 735209
Fax: (502 2) 735207

National Council of Protected Areas
2a Avenida 0-69, Zona 3
Colonia Bran, Guatemala
Phone: (502-2) 518951
Fax: (502-2) 500454

Guatemalan Institute of Tourism
7a Avenida I17, Centro Civico
Zona 4, Guatemala
Phone: (502-2) 311333
Fax: (502 2) 318893

GUINEA

Ministry of Natural Resources, Energy and Environment
B.P. 295, Conakry
Phone: (224) 44 11 86

National Environment Directorate
B.P. 311, Conakry
Phone: (224) 46 10 12

National Directorate of Forests and Hunting
B.P. 624, Conakry
Phone: (224) 44 32 43
Fax: (224) 44 43 87

GUINEA-BISSAU

Ministry of Rural Development and Agriculture
C.P. 71, Bissau
Phone: (245) 212617

GUYANA

Guyana Agency for Health Sciences Education Environment and Food Policy
Greater Georgetown

Guyana Natural Resources Agency
41 Brickdam and Boyle Place
Stabroek, Georgetown
Phone: (592 2) 66549
Fax: (592 2) 7121 1

Guyana Forestry Commission
Younge Street
Kingston

Ministry of Agriculture
Regent and Vlissingen Roads
Georgetown
Phone: (592 2) 69154

Ministry of Health and Public Welfare
Homestretch Avenue
D'Urban Park
Georgetown
Phone: (2) 65861

HAITI

Ministry for Public Works, Transportation and Communication
Palais des Ministeres
B.P. 2002, Port-au-Prince
Phone: (1) 2-0300

Ministry of Agriculture
Directorate of Natural Resources and the Environment
B.P. 1441, Port-au-Prince
Phone: (509 1) 2-1862

Institute for Protection of the National Heritage
B.P. 2484, Port-au-Prince
Phone: (509 1) 2-5286

HONDURAS

Ministry of Natural Resources
Boulevard Miraflores
Tegucigalpa
Phone: (504) 32-7704

Ministry of Planning, Forestry and Fisheries Department
Coordination and Budgeting

Bulevar de las Fuerzas Armadas
Tegucigalpa
Phone: 22-2262

HONG KONG

Environmental Protection Department
24–28th floors, Southern Centre
130 Hennessy Road
Wanchai, Hong King
Phone: (852) 835-1018
Fax: (852) 838-2155

Agriculture and Fisheries Department
Canton Road Government Offices
393 Canton Road, 12th Floor
Kowloon
Phone: (852) 733-2235
Fax: (852) 311-3731

HUNGARY

Ministry for Environment and Regional Policy
Box 351
H-1011 Budapest
Phone: (36 1) 2012964
Fax: (36 1) 201-2846

Institute for Environmental Management
Pf. 369
H-1054 Budapest
Phone: (36 1) 132-9940

ICELAND

Ministry for the Environment
IS-150 Reykjavik
Phone: (354 1) 609600
Fax: (354 1) 624566

Ministry of Culture and Education
Hverfisgata 6
IS-150 Reykjavik

Iceland Nature Conservation Council
P.O. Box 5324
IS-125 Reykjavik
Phone: (354 1) 627855
Fax: (354 1) 627790

INDIA

Ministry of Urban Development
Nirman Bhaman, Rafi Marg
New Delhi 110 001
Phone: (91 11) 3019377

Ministry of Environment and Forests,
B Block Paryavaran Bhavan
CGO Complex
Lodi Road
New Delhi 110003
Phone: (91 11) 306156

Ministry of Agriculture and Rural Development
Krishi Bhavan, Dr.
Rajendra Prasad Road
New Delhi 110001
Phone: (91 11) 382651

Ministry of Health and Family Welfare
Nirman Bhavan
New Delhi
Phone: (91 11) 3018863

Ministry of Ocean Development
Block 12, CGO Complex
Lodi Road
New Delhi
Phone: (91 11) 360874

Ministry of Science and Technology, Technology Bhavan
New Mehrauli Road
New Delhi
Phone: (91 11) 661439

Ministry of Water Resources
Shram Shakti
Bhavan, Rafi Marg
New Delhi 110001
Phone: (91 11) 383098

INDONESIA

Ministry of State for Population and Environment
Jalan Medan Merdeka Barat
15, Jakarta
Phone: (62 21) 371295
Tx: 46143

Department of Home Affairs
J1, Merdeka Utara 7
Jakarta Pusat
Phone: (21) 342222

Ministry of Forestry
Directorate General of Forest Protection and Nature Conservation

8th Floor, Manggala
Wanabhakti Building
Jalan Gatot Subroto
Jakarta
Phone: (62 21) 584818
Fax: (62 21) 251323067

IRAN

Department of the Environment
Ostad Nejat Ollahi, Avenue No. 187
P.O. Box 5181-15875
Teheran
Phone: (98-21) 891261

Research Organization of Agriculture and Natural Resources
P.O. Box 13185-116
Teheran
Phone: (98-21) 944199

IRAQ

Higher Council for Environmental Protection and Improvement
c/o Ministry of Health
P.O. Box 423
Baghdad

Ministry of Agriculture and Irrigation
Baghdad
Phone: (964 1) 8873251

Ministry of Health
Baghdad
Phone: (964 1) 4166156

General Directorate of Preventive Medicine and Environmental Protection
P.O. Box 10062
Baghdad
Phone: (964 1) 4153316

IRELAND

Department of the Environment
Custom House
Dublin
Phone: (353 1) 6793377
Fax: (353 1) 742710

Department of Energy
Clare Street
Dublin
Phone: (353 1) 715233
Fax: (353-1) 773169

Department of the Marine
Leeson Lane
Dublin
Phone: (353-1) 785444
Fax: (353-1) 6r8214

Department of Finance
Office of Public Works
51 St. Stephen's Green
Dublin
Phone: (353-1) 615666
Fax: (353-1) 789527

Department of Agriculture and Food
Kildare Street
Dublin
Phone: (353-1) 78901 1
Fax: (353-1) 616263

Department of Foreign Affairs
80 St. Stephen's Green
Dublin
Phone: (353-1) 780822
Fax: (353-1) 717149

ISRAEL

Ministry of the Environment
P.O. Box 6234
Jerusalem 91061
Phone: (972-2) 701411
Fax: (972-2) 385038

Ministry of Agriculture
Hakirya, 8 Arania Street
Tel Aviv 61070
Phone: (972-3) 21121 1
Fax: (972-3) 268899

Land Development Authority
Forest Department
P.O. Box 45
26013 Kiryat-Hayim
Haifa 32000
Phone: (972 4) 411983
Fax: (972 4) 411971

Ministry of Health
Public Health Services
Haarbaah 14
Tel Aviv 64739
Phone: (972-3) 5634707

Ministry of Foreign Affairs
Division for International Cooperation
Mendler 7, Hakirya
Tel Aviv 61070
Phone: (972-3) 219271

Nature Reserves Authority
Yirmeyahu 78
Jerusalem 94467
Phone: (972 2) 387471

Jewish Agency, Settlement Department
Kaplan 17
Tel Aviv 64734
Phone: (972-3) 5423423

Palestinian Centre for the Environment
P.O. Box 897
Jerusalem 91007
Phone: (972-2) 854670
Fax: (972-2) 283336

ITALY

Ministry of the Environment
Piazza Venezia 11
1-00187 Rome
Phone: (39-6) 675 931
Fax: (39-6) 675 93 297

Ministry of Foreign Affairs
Department of Development Cooperation
Piazzale delle Farnesina 1
I-00194 Rome
Phone: (39 6) 3691 1

Ministry of Cultural Heritage and Environment
9 Via del Collegio Romano 27
1-00186 Rome
Phone: (39 6) 6723

JAMAICA

Ministry of Foreign Affairs
85 Knutsford Boulevard
Kingston 5
Phone: (809) 926-4220
Fax: (809) 929-6733

Ministry of Finance, Development Planning and Production
Includes: Natural Resources Conservation Department
P.O. Box 305
53½ Molynes Road
Kingston 10
Phone: (1-809) 923-5155

Ministry of Construction
2 Hagley Park Road
Kingston 10
Phone: (809) 926-1590

Department of Forestry and Soil Conservation
4 Hillman Road
Kingston 8
Phone: (809) 924-2667

Ministry of Health
Environmental Control Division
Life of Jamaica Building
61 Half Way Tree Road
Kingston 10
Phone: (809) 929-6463

Jamaica Conservation and Development Trust
P.O. Box 1225
Kingston S
Phone: (I 809) 922-2217
Fax: (809) 922-0665

JAPAN

Environment Agency of Japan
3-1-1 Kasumigaseki, Chiyada-ku
Tokyo 100
Phone: (81-3) 3581-3351

National Institute for Environmental Studies
16-2 Onogawa, Tsukuba
Ibaraki 305
Phone: (81-298) 516111
Fax: (81-298) 58-2645

Ministry of Health and Welfare
2-2-1 Kasumigaseki, Chiyada-ku
Tokyo 100
Phone: (81-3) 35014967

Ministry of Foreign Affairs
2-2-1 Kasumigaseki, Chiyada-ku
Tokyo 100
Phone: (81 3) 3581-3882

Ministry of International Trade and Industries
Environmental Policy Division
1-3-1, Kasumigaseki, Chiyada-ku
Tokyo 100
Phone: (81-3) 3501-1511

Japan Meteorological Agency
Chiyada-ku
Tokyo 100

Phone: (81 3) 3211-7084

JORDAN

Ministry of Municipal and Rural Affairs and the Environment
Department of the Environment
P.O. Box 1799
Amman
Phone: (962 6) 634192

Ministry of Agriculture
Department of Forests and Soil Conservation
P.O. Box 2179
Amman
Phone: (962-6) 842751

Ministry of Water Resources and Irrigation
Water Authority
P.O. Box 2412
Amman
Phone: (962-6) 666111

KAZAKHSTAN

Ministry of Ecology and Bioresources
Alma-Ata
Phone: (3272) 33-46-11

Chamber of Commerce and Industry
Prospekt. Ablaikhana 93/95
480091 Alma-Ata
Phone: (3271) 62-14-46
Fax: (3272) 62-05-95

KENYA

Ministry of Environment and Natural Resources
P.O. Box 67839
Nairobi
Phone: (254-2) 332383

Forest Department
P.O. Box 30513
Nairobi
Phone: (254-2) 229261
Fax: (254-2) 340260

Ministry of Tourism and Wildlife
P.O. Box 30027
Nairobi
Phone: (254-2) 891601

Kenya Wildlife Service
P.O. Box 40241
Nairobi
Phone: (254-2) 501084
Fax: (254-2) 505866

Permanent Presidential Commission on Soil Conservation and Afforestation
Office of the President
P.O. Box 30510
Nairobi
Phone: (254-2) 21034

KIRIBATI

Ministry of Natural Resource Development
P.O. Box 64
Bairiki
Tarawa
Phone: (686) 21099
Fax: (686) 21120

Ministry of Health and Family Planning
P.O. Box 268
Bikenibeu
Tarawa
Phone: (686) 28081

KOREA, DEMOCRATIC PEOPLE'S REPUBLIC OF (NORTH KOREA)

Korea Chemicals Exports
Pyongyang
Phone: 45477

Ministry of Forestry
Pyongyang

Ministry of Natural Resources Development
Pyongyang

Ministry of Public Health
Pyongyang

Fisheries Commission
Pyongyang

KOREA, REPUBLIC OF (SOUTH KOREA)

Ministry of Environment
7-16 Sinchon-dong Songpa-gu
Seoul 138-240
Phone: (82-2) 421-0301

Ministry of Agriculture and Fisheries
Chungang-dong, Gwacheon-shi
Gyonggi-do 427-760
Phone: (82-2) 503-7208

Ministry of Construction
Land Planning Bureau
1 Chungang-dong, Gwacheon-shi
Gyonggi-do 427-760
Phone: (82 2) 503-7312

Ministry of Health and Social Affairs
Chungang-dong, Gwacheon-shi
Gyonggi-do 427-760
Phone: (82-2) 503-7504

Ministry of Home Affairs
Subsection of Nature Preservation
Government United Building
77 Sejong-ro, Chongno-gu
Seoul 110-760
Phone: (82-2) 731-2350

KUWAIT

Environment Protection Council
P.O. Box 24395
13104 Safat
Phone: (965) 245 6835
Fax: (965) 242 1993

Ministry of Public Health
Environmental Protection Department
P.O. Box 35035

Public Authority for Agricultural Affairs and Fish Resources
P.O. Box 21422
10375 Safat
Phone: (965) 476 2111
Fax: (965) 476 5551

Kuwait Fund for Arab Economic Development
P.O. Box 2921
Kuwait
Phone: (965) 43 9075

Kuwait Environment Protection Society
c/o ROPME
P.O. Box 26388
13124 Safat
Phone: (965) 531 2140
Fax: (965) 532 4172

KYRGYZSTAN

Office of the President
ul. Kirova 205, Bishbek
Kyrgyzstan Society for Nature Protection
Pervomajskaya 54
720000 Bishbek
Phone: 22 15 91

LAOS

Ministry of Agriculture and Forestry
Department of Forestry and Environment
Vientiane

Wildlife and Fisheries Conservation Division
Environmental Protection Division

Ministry of Public Health
Vientiane
Phone: 2223

MALAWI

Department of Research and Environmental Affairs
Private Bag 388
Lilongwe 3
Phone: (265) 722780

Department of Forestry
P.O. Box 30048
Lilongwe 3
Phone: (265) 731322

Department of National Parks and Wildlife
P.O. Box 30131
Lilongwe 3
Phone: (265) 723566
Fax: (265) 723089

MALAYSIA

Ministry of Science, Technology, and the Environment
Wisma Sime Darby, 14th Floor
Jalan Raja Laut
50662 Kuala Lumpur
Phone: (60-3) 293 8955
Fax: (60-3) 293 6006

Department of Wildlife and National Parks

West Malaysia
Km 10 Jalan Cheras
50664 Kuala Lumpur
Phone: (60 3) 905 2872
Fax: (60 2) 905 2873

Forestry Department
Jalan Sultan Salahuddin
50600 Kuala Lumpur

Ministry of National and Rural Development
Bangunan Kipleks Kewangan
5th–10th Floors
Jalan Raja Chulan
50606 Kuala Lumpur
Phone: (3) 2912622
Fax: (3) 22611339

MALDIVES

Ministry of Planning and Environment
Fax: (960) 32735 1

Ministry of Fisheries and Agriculture
Ghaazee Building
Ameer Ahmed Magu, Malé 20-05
Phone: (960) 322625
Fax: (960) 326558

Ministry of Health and Welfare
Ghaazee Building, Ameer Magu, Male 20-05
Phone: (960) 323216

MALI

Ministry of Environment Resources and Livestock
B.P. 275, Bamako
Phone: (223) 225850

Ministry of Health, Solidarity and Elderly
Koulouba, Bamako
Phone: 22 55 01

MALTA

Ministry of Education and the Interior, Environmental Division
Floriana
Phone: 23-12-93

Department of Agriculture and Fisheries
14 M.A. Vassalli Street, Valletta
Phone: (356) 22 49 41

MARSHALL ISLANDS

Environmental Protection Authority
P.O. Box 1322
Majuro, Marshall Islands 96960

Ministry of Resources and Development
Cabinet Building
P.O. Box 326
Majuro, Marshall Islands 96960
Phone: (692 9) 3206

MARTINIQUE

Préfecture, Rue Victor-Severe
97262 Fort-de-France Cadex
Phone: (596) 63 18 61
Fax: (596) 7140 29

Ministry of Overseas Departments and Territories
27 rue Oudinot
F-75700 Paris 07 SP, France
Phone: (I 33) 47 83 01 23

MAURITANIA

Ministry of Rural Development
Directorate of Nature Protection, Reforestation, and Wildlife
B.P. 170, Nouakchott
Phone: (222-2) 51763
Fax: (222-2) 51834

Ministry of Water Resources and Energy
B.P. 12559 Nouakchott

MAURITUS

Ministry of Environment and Land Use
Edith Cavell Street
Port Louis

Ministry of Agriculture, Fisheries, and Natural Resources
4th Floor, Government House
Port Louis

Council for Development, Environmental Studies, and Conservation
P.O. Box 1124
Port Louis

Mauritius Council for Environmental Studies, Conservation, and Youth Agenda
G.P.O. Box 304
Port Louis
Phone: (230) 2165593
Fax: (230) 2087882
(Broad interests.)

Ministry of Local Government
Government Center
Port Louis
Phone: 2011215

MEXICO

Secretariat of Urban Development and Ecology
Avenida Constituyentes 947
Edificio B. P.A.
Colonia Belam de las Flores
0 1 1 1 0 Mexico, D.F.
Phone: (52 5) 271-8481 or 271-0355
Fax: (52 5) 271-6614

Secretariat of Agriculture and Water Resources
Insurgentes Sur 476
13.Q. Piso
06038 Mexico, D.F.
Phone: (52 5) 584-0066

National Forestry and Agricultural Research Institute
Apartado Postal 6-882
06600 Mexico, D.F.
Phone: (52 5) 687-7451

Secretariat of Fisheries
Avenida Alvaro Obregen 269
611 Piso
06700 Mexico, D.F.
Phone: (52 5) 511-9278

National Forestry Commission
Avenida Mexico 190
Coyoacdn
04100 Mexico, D.F.
Phone: (52 5) 534-9707 or 5247862

MICRONESIA, FEDERATED STATES OF

Department of Human Resources
Environmental Protection Board
P.O. Box 490, Kolonia
Eastern Caroline Islands 96941
Phone: (691) 3202646

MONACO

Public Works Department
B.P. 522
MC-98015 Monaco Cedex
Phone: (93) 15-18-00
Fax: (93) 15-92-33

MONGOLIA

Ministry of Fuel, Power and Energy
Bagatoriuu 6
Ulan Bator 46
Phone: (1) 328350
Fax: (1) 341430

Ministry of Nature and Environment
Ulan Bator
Phone: (1) 321401

MONTSERRAT

Ministry of Agriculture, Trade, Lands, and Housing
Plymouth
Phone: 2546

Ministry of Education, Health, and Community Services
Plymouth
Phone: 3321

MOROCCO

Ministry of the Interior
General Directorate of Urbanism, Territorial Management and the Environment
Rabat
Phone: (212-7) 60262

Division of the Environment
Phone: (212 7) 63357
Fax: (212 7) 68426

Ministry of Agriculture and Agrarian Reform
51, Avenue de France
Rabat
Phone: (212-7) 760993

Directorate of Water and Forests and Soil Conservation
B.P. Rabat
Phone: (212-7) 625-65

Ministry of Health

335 Avenue Muhammad V,
Rabat
Phone: (212-7) 61121

MOZAMBIQUE

Ministry of Health
Eduardo Mondlane
Caixa Postal 264
Maputo
Phone: (1) 427131
Fax: (1) 432103

Ministry of Mineral Resources
Avenida Fernao Magalhaes 34
Caixa Postal 2203
Maputo
Phone: (1) 429615

Ministry of Agriculture
National Directorate of Forestry and Wildlife
Caixa Postal 1406
Maputo
Phone: (258 1) 460129
Fax: (258 1) 460060

Ministry of Construction and Water Affairs
National Water Directorate
Caixa Postal 268
Maputo

MYANMAR

Ministry of National Planning and Economic Development
228-240 Strand Road
Yangon
Phone: (1) 57316

Ministry of Foreign Affairs
National Commission for Environmental Affairs
Prome Court, Prome Road
Yangon
Phone: (1) 83333

NAMIBIA

Ministry of Wildlife, Conservation, and Tourism
Directorate of Wildlife, Conservation and Research
Private Bag 13306
Windhoek 9000
Phone: (264-61) 63131
Fax: (264-61) 63195

Ministry of Agriculture, Water and Rural Development
Private Bag 13191
Windhoek Directorate of Forestry
Private Bag 13184
Windhoek

Ministry of Fisheries and Marine Resources
Private Bag 13355
Windhoek

Ministry of Mines and Energy
Private Bag 13297
Windhoek

National Planning Commission
Private Bag 13356
Windhoek

Other Organizations

Desert Ecological Research Unit of Namibia
P.O. Box 953
Walvis Bay 9190

Namibia Nature Foundation
P.O. Box 245
Windhoek 9000

Wildlife Society of Namibia
P.O. Box 3508
Windhoek

NAURU

Ministry for Island Development and Industry
Yaren
Phone: (674) 3320

Ministry for Health and Education
Yaren
Phone: (674) 3350

Department of Island Development and Industry
Nauru
Phone: (674) 3320

KINGDOM OF NEPAL

Ministry of Forestry Development and Environment
Singh Durbar
Kathmandu

Phone: (977 1) 220160

National Planning Commission
Environment and Conservation Division
P.O. Box 1284
Kathmandu
Phone: (977 1) 228200, ext. 252
Fax: (977 1) 226500

Ministry of Agriculture
Singh Durbar
Kathmandu
Phone: (977 1) 225108

Other Organizations

IUCN—The World Conservation Union
Nepal Office
P.O. Box 3923
Kathmandu
Phone: (977 1) 21506
Fax: (977 1) 226820

King Mahendra Trust for Nature Conservation
Babar Mahal
P.O. Box 3712
Kathmandu
Phone: (977 1) 223229
Fax: (977 1) 226602

Nepal Forestry Association
P.O. Box 2761
Babar Mahal
Kathmandu

Royal Nepal Academy of Science and Technology
P.O. Box 3323
New Baneswar
Kathmandu
Phone: 213060

THE KINGDOM OF NETHERLANDS

Ministry of Housing, Physical Planning, and the Environment
Directorate General for Environment
Postbus 450
NL-2260 MB Leidschendam
Phone: (31 70) 317-4174
Fax: (31 70) 317-4831

Ministry of Agriculture, Nature Management, and Fisheries
Bezuidenhoutseweg 73
The Hague
Phone: (31 70) 379-391 1

Ministry of Foreign Affairs
Postbus 20061
NL-2500 EB
The Hague
Phone: (31 70) 348-6486
Fax: (31 70) 348-4848

Other Organizations

Environment and Development Service for NGO
Damrak 28-30
NL-1012 LJ Amsterdam
Phone: (31 20) 623-0823
Fax: (31 20) 620-8049

Information Centre for Low External Input and Sustainable Agriculture
Postbus 64
NL3830 AB Leusden
Phone: (31 33) 943086
Fax: (31 33) 940791

Institute for Nature Conservation Education
Postbus 20123
NL-1000 HC Amsterdam
Phone: (31 20) 622-8115

Council for Nature Conservation
Maliebaan 12
NL-3581 CN Utrecht
Phone: (31 30) 331441

Leiden University
Center for Environmental Studies
Postbus 9518
NL-2300 RA Leiden
Phone: (31 71) 277-474
Fax: (31 71) 277-496

NEW ZEALAND

Ministry of Health
P.O. Box 5013
Wellington
Phone: (4) 496 2000
Fax: (4) 496 2340

Ministry of Environment
P.O. Box 10-362
Wellington
Phone: (64-4) 473 4090
Fax: (64-4) 471 0195

Department of Conservation
P.O. Box 10-420
Wellington
Phone: (64-4) 471-0726
Fax: (64-4) 471-1082

Ministry of Agriculture and Fisheries
P.O. Box 297
Wellington
Phone: (64 4) 486-1029

Ministry of External Relations and Trade
Private Bag
Wellington
Phone: (64-4) 472-8877
Fax: (64-4) 472-9596

Department of Scientific and Industrial Research
P.O. Box 1578
Wellington
Phone: (64 4) 472-9979
Fax: (64 4) 472-4025

Environment and Conservation Organizations of New Zealand
P.O. Box 11-057
Wellington 6034
Phone/Fax: (64 4) 384-6971

Royal Society of New Zealand
P.O. Box 598
Wellington
Phone: (64 4) 472-7421
Fax: (64 4) 473-1841

NICARAGUA

Ministry of Natural Resources
KM 12 ½ Carreta Norte
Apdo 5123
Managua JR
Phone: (2) 31110
Fax: (2) 31274

Ministry of Government
Apdo 68
Managua
Phone: (2) 27531

Nicaraguan Institute of Natural Resources and Environment
Apartado 5123
Managua
Phone: (505 2) 3-1110
Fax: (505 2) 31274

NIGER

Ministry of Water Resources and the Environment
B.P. 578, Niamey
Phone: (227) 73 33 29

Directorate of the Environment
B.P. 578
Niamey
Phone: (227) 73 33 29

Directorate of Wildlife, Fisheries and Fish Farming
B.P. 721, Naimey
Phone: (227) 73 49 63
Fax: (227) 73 22 15

Directorate of Hygiene and Sanitation
B.P. 371, Niamey
Phone: (227) 72 37

NIGERIA

Federal Ministry of Works and Housing
P.M.B. 12698, Lagos
Phone: (234-1) 682625

Federal Ministry of Agriculture and Natural Resources
P.M.B. 135, Garki, Abuja
Phone: (234-9) 5230177

Federal Department of Forestry and Agricultural Land Resources
P.M.B. 12613, Ikoye, Lagos
Phone: (234 1) 684178

Federal Ministry of Science and Technology
P.M.B. 5054, Ibadan
Phone: (234 22) 414441

Campaign for Environmental Protection Awareness
P.O. Box 1066
Ikeja-Lagos

Nigerian Conservation Foundation
P.O. Box 74638

Victoria Island, Lagos
Phone: (234 1) 686163
Fax: (234 1) 685378

NORWAY

Royal Ministry of Environment
Myngtgaten 2
Postboks 8013-Dep.
N-0030 Oslo 1

Center for International Climate and Energy Research in Oslo
P.O. Box 1066, Blindern
N-0316 Oslo
Phone: (47 2) 854286
Fax: (74 2) 856284

OMAN

Ministry of Regional Municipalities and Environment
P.O. Box 3461
Muscat
Phone: (968) 696444
Fax: (968) 602320

Ministry of Health
Environmental Health Section
P.O. Box 393
Muscat

Ministry of Agriculture and Fisheries
P.O. Box 467
Muscat
Phone: (968) 696312

Ministry of Commerce and Industry
P.O. Box 550
Muscat
Phone: (1) 799500

PAKISTAN

Ministry of Planning and Development
Islamabad
Phone: (51) 826366

Ministry of Food, Agriculture, and Cooperatives
National Council for Conservation of Wildlife in Pakistan
485 Street 84, Sector G-6/4
Islamabad
Phone: (92 51) 823520

Pakistan Agricultural Research Council
Arid Zone Research Institute
Brewery Road Quetta
Baluchistan
Phone: (928 1) 75006

Ministry of Housing and Works
Block B, Pakistan Secretariat
Islamabad
Phone: (92 51) 813373

Pakistan Forest Institute
P.O. Forest Institute Campus
Peshawar
Phone: (92-521) 41388

Energy and Environment Society of Pakistan
Model Town, Lahore 54700
Phone: (92 42) 851583

Environmental Management Society
141-A SMCH Society
Karachi 74400
Phone: (02 21) 446644

Environmental Protection Society of Pakistan
2nd Floor, PAAF Building
7D Kashmir Egerton Road
Lahore
Phone: (92-42) 360800
Fax: (92-42) 303825

Pakistan Institute for Environment Development Action Research
G 41 Bhitai Road, F7/1
Islamabad
Phone: (92-51) 823790

PANAMA

Ministry of Planning and Economic Policy
National Environment Commission
Apartado 2694, Zona 3
Panama
Phone: (507) 32-6055

Ministry of Health
Apartado 2048
Panama 1
Phone: (507) 25-6080
Fax: (507) 257-5276

Institute of Renewable Natural Resources
Apartado 2016
Paraiso
Phone: (507) 32-4518

Fax: (507) 32-4975

PAPUA NEW GUINEA

Ministry for Environment and Conservation
Central Government Offices
P.O. Office, Wards Strip
Waigani
Phone: (675) 271788
Fax: (675) 213826

University of Papua New Guinea
Department of Environment Sciences
Box 320, University P.O.
Phone: (675) 245228

PARAGUAY

Ministry of Public Health and Social Welfare
Avenida Pettirossi y Brasil
Asuncion
Phone: (21) 207-328

PERU

Ministry of Agriculture
Avenida Salaverry s/n
Edificio Ministerio de Trabajo
Jesus Marla, Lima
Phone: (51-14) 323629
Fax: (14) 320990

Association of Ecology and Conservation
Avenida Dos de Mayo 527
Lima 18
Phone: (51-14) 472369

PHILIPPINES

Department of Environment and Natural Resources
DENR Building
Visayas Avenue
Diliman, Quezon City
Phone: (2) 97 66 26

Department of Agriculture
Elliptical Road
Diliman, Quezon City
Phone: (63 2) 998-741

Department of Health
San Lazaro Hospital Compound
Rizal Avenue
Santa Cruz, Metro Manila
Phone: (63 2) 266-806

National Water Resources Council
NIA Building
EDSA, 8th Floor
Quezon City
Phone: (632) 952-603

Ecological Society of the Philippines
P.O. Box 17499 MCC
Metro Manila
Phone: (63 2) 631-7351
Fax: (63 2) 631-7357

Foundation for Sustainable Development, Inc.
L 25 K-J Street
Kamias
Quezon City, Metro Manila
Phone: (63 2) 922-6397

POLAND

Ministry of Environmental Protection, Natural Resources, and Forestry
Wawelska 52.54
PL-02-067 Warsaw
Phone: (48 22) 25 11 33
Fax: (48 22) 25 33 55

Economic Academy in Krakow
Center for Industrial Development and Environmental Protection
Rakowiecka 27
PL-31-510 Krakow
Phone: (48 12) 21 00 99, ext. 329 or 332
Fax: (48 12) 2121 82

Environment and Development
Krucza 511 19
PL-00-548 Warsaw
Phone: (48 22) 21 36 70
Fax: (48 22) 29 52 63

Foundation for Ecology and Health
Ujazdowskie 13
PL-00-567 Warsaw
Phone: (48 22) 694 2153
Fax: (48 22) 694 16 48

Institute for Sustainable Development
Krzywickiego 9
PL-02-078 Warsaw
Phone: (48 22) 25 2558

Fax: (49 22) 25 34 61

PORTUGAL

Ministry of Environment and Natural Resources
Rua de 0 Seculo 51
P-1200 Lisbon
Phone: (351 1) 346-2751
Fax: (351 1) 346-8469

PUERTO RICO

Department of Natural Resources
P.O. Box 5887
Puerta de Tierra Station
San Juan, PR 00906 USA
Phone: (809) 724-8774

Department of Agriculture
P.O. Box 10163
Santurce
San Juan PR 00908 USA
Phone: (809) 721-2120

Conservation Trust of Puerto Rico
P.O. Box 4747
San Juan, PR 00902 USA
Phone: (809) 722-5834

QATAR

Environmental Protection Committee
P.O. Box 7634
Doha
Phone: (974) 320825
Fax: (974) 415246

Ministry of Municipal Affairs and Agriculture
P.O. Box 2727
Doha
Phone: (974) 413535
Fax: (974) 414868

Ministry of Health
P.O. Box 42
Doha
Phone: (974) 324566

RUSSIA

Ministry of Ecology and Natural Resources
Moscow
Phone: (095) 252-23-05

RWANDA

Ministry of Planning and International Cooperation
B.P. 46, Kigali
Phone: 75513

Ministry of Environment
Kigali

SAINT KITTS-NEVIS

Ministry of Natural Resources and Environment
Administration Building
Charlestown, Nevis
Phone: (809) 469-5374

SAINT LUCIA

Ministry of Health and Local Government
Chaussee Road
Castries
Phone: (809) 452-2827

SAINT VINCENT AND THE GRENADINES

Ministry of Health and Environment
Kingstown
Phone: (809) 456-1111, ext. 462

SAN MARINO

Ministry of Territory, Environment and Agriculture
47031 San Marino
Phone: (549) 992345

SAO TOME AND PRINCIPE

Ministry of Environment and Social Affairs
C.P. 130, Sã Tomé
Phone: (12) 22824

SAUDI ARABIA

Ministry of Agriculture and Water
Airport Road
Riyadh 1195
Phone: (1) 401-6666

SENEGAL

Ministry of Environment and Nature Protection
Dakar
Phone: 214201

Fax: 238488

SEYCHELLES

Department of Environment
P.O. Box 445
Victoria, Mahé
Phone: (248) 22881

Ministry of Agriculture and Fisheries
Independence House
P.O. Box 92
Victoria, Mahé
Phone: (248) 25333
Fax: (248) 25245

SIERRA LEONE

Department of Agriculture and Forestry
Youyi Building, 1st Floor
Brookfields, Freetown
Phone: 232 (22) 41500, ext. 516

SINGAPORE

Ministry of Environment
Environment Building
40 Scotts Road
Singapore 0922
Phone: (65) 7327733
Fax: (65) 7319456

SLOVAKIA

Ministry of Environment
Hlboka 2
812 35 Bratislava
Phone: (7) 492-451
Fax: (7) 311-368

SLOVENIA

Ministry of Environment
Zupanciceva St. 6
61000 Ljubljana
Phone: (61) 154208
Fax: (61) 224548

SOLOMON ISLANDS

Ministry of Natural Resources
P.O. Box G24
Honiara
Phone: (677) 22944

SOMALIA

Ministry of Health
Mogadishu
Phone: (252) 31055

Ministry of Forestry
P.O. Box 1759
Mogadishu, Somalia
Tx: 736

Ministry of Minerals and Water Resources
Mogadishu
Phone: 80980

SOUTH AFRICA

Ministry of Environment and Water Affairs
Private Bag X313
Pretoria 0001
Phone: 27 (12) 299-3391
Fax: 27 (12) 284-254

SPAIN

Ministry of Industry and Energy
Paseo de la Castellana, 160
28071 Madrid
Phone: (1) 349-4000
Fax: (1) 457-8066

SRI LANKA

Ministry of Environment
6th Floor
Unity Plaza Building
Colombo 4
Phone: (1) 58874

SUDAN

Ministry of Industry and Commerce
P.O. Box 2184
Khartoum
Phone: (249 11) 72569

Ministry of Health
P.O. Box 287
Khartoum
Phone: 73000

SURINAME

Ministry of Natural Resources and Energy
Mr. Dr. J. C. de Mirandastraat 13-15

Paramaribo
Phone: (579) 474666
Fax: (579) 472911

SWAZILAND

Ministry of Health
P.O. Box 5
Mbabane
Phone: (268) 42431

Ministry of Natural Resources and Energy
P.O. Box 100
Lobamba
Phone: (268) 61151

Ministry of Natural Resources
P.O. Box 57
Mbabane
Phone: 448708

SWEDEN

Ministry of Environment
Tegelbacken 2
103 33 Stockholm
Phone: (8) 7631000
Fax: (8) 241629

SWITZERLAND

Chancery of Swiss Confederation
Bundeshaus West
CH-3003 Berne
Phone: (41) (31) 3222111

SYRIA

Ministry of Health
Parliament Street
Damascus
Phone: (96311) 233801

TAIWAN

Department of Health
100 Aikuo E. Road
Taipei
Phone: (886) (2) 3210151
Fax: (886) (2) 312 2907

TANZANIA

Ministry of Health and Social Welfare
P.O. Box 9083
Dar es Salaam
Phone: (255) (51) 20261

THAILAND

Ministry of Environment and Technology
Thanon Phra Ram Hok
Phyaya Thai, Bangkok 10400
Phone: (2) 245-7799
Fax: (2) 245-8636

TOGO

Ministry of Rural Development and Environment
Avenue de Sarawaka
Lome
Phone: (228) 215671

TONGA

Ministry of Natural Resources
Administration Building
Vuna 14
Phone: (676) 22-655

Ministry of Health
P.O. Box 59
Nuku'alofa, Tonga
Phone: (676) 21-200

TRINIDAD AND TOBAGO

Ministry of Health
35-37 Sackville Street
Port of Spain
Phone: (809) 675-5244
Fax: (809) 638-8620

TUNISIA

Ministry of the Environment and Territorial Management
National Agency for Environmental Protection
15 Rue 800, Montplaisir
1002 Tunis-Belvédere
Phone: (216-1) 785618
Fax: (216-1) 789-844

Ministry of Education and Scientific Research
National Institute of Scientific and Technological Research
B.P. 95, Hammam-Lif
Phone: (216 3) 291-044

Tunisian Association for Protection of Nature and the Environment
12 Rue Tantaoui el Jawhari
El Omrane, Tunis 1005

Phone: (216 1) 288-141
Fax: (216 1) 797-295

TURKEY

Ministry of Environment
Atatfirk Bulvari 143
Bakanliklar-Ankara
Phone: (90 4) 419 15-53
Fax: (90 4) 417 79 71

Society for the Protection of Wildlife
P.K. 18, Bebek
80810 Istanbul
Phone: (90-1) 163 63 24
Fax: (90-1) 163 05

Environmental Problems Foundation of Turkey
Kenedi Caddesi 33/7
06660 Kavaklidere-Ankara
Phone: (90 4) 425 55 08
Fax: (90 4) 118 51 18

Turkish Foundation for Protection of Wildlife
Istanbul
Phone: (90-1) 126 11 00

Turkish Association for the Conservation of Nature and Natural Resources
Menekse Sokak 29/4, Kizilay
06440 Ankara
Phone: (90-4) 425 19 44
Fax: (90-4) 417 95 52

TURKMENIA

Academy of Sciences of Turkmenia
Desert Institute
Gogol Street 15
744000 Ashgabat

Ministry of Natural Resources
Cockburn Town
Grand Turk
Phone: (1-809) 946-2801

TUVALU

Ministry for Natural Resources and Home Affairs
Vaiaku, Funafuti

UGANDA

Ministry of Environment Protection
P.O. Box 9629
Kampala
Phone: (256-41) 257976

Ministry of Agriculture and Forestry
Kampala

Ministry of Wildlife and Tourism
Game Department
P.O. Box 4
Entebbe
Phone: (256 41) 20073

Uganda National Parks
P.O. Box 3530
Kampala

Joint Energy and Environment Projects
P.O. Box 1684
Jinja
Phone: (256 43) 20054

Makerere University
Faculty of Agriculture and Forestry
P.O. Box 7062
Kampala
Phone: (256 41) 242271

Uganda Environmental Law Association
P.O. Box 1188
Kampala

Wildlife Clubs of Uganda
P.O. Box 4596
Kampala
Phone: (256 41) 256534

UKRAINE

Ministry of Environmental Protection
Kreshchatik Street 5
252001 Kiev
Phone: (7-044) 228-0644
Fax: (7-044) 229-8383

Ministry of Forestry
Kreshchatik Street 5
252001
Kiev
Phone: (7-044) 226-3253
Fax: (7-044) 4160317

Ministry of Assistance for the Human Consequences of the Chernobyl Accident
Lesi Ukrainki Square 1
252196 Kiev
Phone: (7-044) 296-8395

Fax: (7-044) 294-7796

State Committee for Water Resources
Chervonoarmiyska 8
252601 Kiev
Phone: (7 044) 226-2607

State Committee for Geology and Mineral Resource Usage
Kiev
Fax: (7-044) 228-6221

UNITED ARAB EMIRATES

Higher Environmental Committee
c/o Ministry of Health
P.O. Box 1853
Dubai
Phone: (971 4) 23 30, 21
Fax: (971 4) 21 42-98

Ministry of Agriculture and Fisheries
P.O. Box 1509
Dubai

Ministry of Electricity and Water Resources
P.O. Box 629
Dubai

Abu Dhabi Fund for Arab Economic Development
P.O. Box 814
Abu Dhabi
Phone: (971-) 82 2865

UNITED KINGDOM

Ministry of Agriculture, Fisheries, and Food
Whitehall Place
London SWIA 2HH
Phone: (44 71) 270 3000

Department of Agriculture and Fisheries for Scotland
Pentland House, 47 Robb's Loan
Edinburgh EH4 ITW
Phone: (44 31) 556 8400

Department of Agriculture for Northern Ireland
Dundonald House
Upper Newtownards Road
Belfast
Northern Ireland BT4 3SB
Phone: (44 232) 650111

Department of Energy
1 Palace Street
London SWIE 5HE
Phone: (44 71) 238 3000

Department of the Environment
2 Marsham Street
London SWIP 3EB
Phone: (44 71) 276 3000

Welsh Development Agency
Treforest Industrial Estate
Pontypridd
Wales CF37 5UT
Phone: (44 44385) 2666

National Radiological Protection Board
Chilton, Didcot, Oxon
OXI I ORQ
Phone: (44 235) 831600

Overseas Development Administration
Eland House, Stag Place
London SWIE 5DH
Phone: (44 71) 273 3000

Royal Botanic Gardens
Kew, Richmond
Surrey TW9 3AB
Phone: (44 81) 940 1171

Royal Commission on Environmental Pollution
Church House
Great Smith Street
London SWIP 3BL
Phone: (44 71) 276 2080

URUGUAY

Ministry of Health
Avenida 18 de Julio 1892
11200 Montevideo
Phone: (2) 400101
Fax: (2) 485360

UZBEKISTAN

Ministry of Health
ul. Navio 12
Tashkent
Phone: (3712) 441202

VANUATU

Ministry of Lands and Natural Resources
P.O. Box 151
Port Vila
Phone: 23105

VENEZUELA

Ministry of Environmental Affairs and Renewable Natural Resources
Torre Norte
Centro Simon Bolivar
Caracas 1010
Phone: (2) 4081111

VIETNAM

Ministry of Science, Technology and Environment
39 Tran Hung Dao
Hanoi
Phone: (84) (4) 63379

WESTERN SAMOA

Department of Lands and Environment
Private Bag
Apia
Phone: (685) 22481
Fax: (685) 21504

YEMEN

Ministry of Fish Resources
P.O. Box 19179
Sanaa
Phone: (1) 262866
Fax: (1) 263165

Ministry of Health
Altahrir Square
Sanaa
Phone: (1) 252222
Fax: (1) 244143

YUGOSLAVIA

Ministry of Environment
Belgrade
Phone: (11) 602555

ZAIRE

Ministry of Public Health and Social Welfare
B.P. 3088
Kinshasa-Gombe
Phone: (243) 31750

ZAMBIA

Ministry of Health
Woodgate House
1st and 2nd Floors, Cairo Road
P.O. Box 30205
Lusaka
Phone: (260) (1) 211528

ZIMBABWE

Ministry of Environment and Tourism
Private Bag 7753
Karigamombe Centre, 14th Floor
Causeway
Harare
Phone: (263) (4) 794455

APPENDIX III

INTERNATIONAL INTERGOVERNMENTAL AGENCIES AND DONOR ASSISTANCE ORGANIZATIONS

African Development Bank (ADB)
B. P. 1387, Abidjan 015, Côte d'lvoire
Phone: (224) 20 44 44
Tx: 23717
C: AFDEV
Fax: (225) 21 74 71

African Ministerial Conference on the Environmental (AMCEN)
The Secretariat of the Conference
c/o UNEP Regional Office for Africa
P.O. Box 30552
Nairobi, Kenya
Phone: (254 2) 521840
Tx: 22068
Fax: (254 2) 520711

African Timber Organization (ATO)
B. P. 1077
Libreville, Gabon
Phone: (241) 73 29 28
Tx: oab 5620 go

Arab Bank for Economic Development in Africa (ABEDA)
P.O. Box 2640
Khartoum, Sudan
Phone: (249 1 1) 73646
Tx: 22248 sd.
C: BADEA
Fax: (249 1 1) 70600

Arab Center for the Studies of Arid Zones and Dry Lands (ACSAD)
P.O. Box 2440
Damascus, Syria
Phone: (963 11) 755713
C: ACSAD
Tx: ACSAD 412697

Arab Fund for Economic and Social Development (AFESD)
P.O. Box 21923
Safat 13080
Kuwait
Phone: (965) 2451580
Tx: 22153 kt.
C: INMARABI KUWAIT
Fax: (965) 2416758

Arab League Educational, Cultural and Scientific Organization (ALECSO)
B.P. 1120

Tunis, Tunisia.
Phone: (216 1) 78 44 66
Tx: 13825 tn
Fax: (216 1) 78 49 65

Caribbean Community (CARICOM)
P.O. Box 1113
Port of Spain
Trinidad and Tobago
Phone: (1-809) 623-2280
Fax: (1-809) 6238485

Caribbean Environmental Health Institute (CEHI)
P.O. Box III 1
the Morne, Castries
St. Lucia
Phone: (1-809) 452-2501
Fax: (1-809) 4532721

Caribbean Development Bank (CDB)
P.O. Box 408
Wildey, St. Michael
Barbados
Phone: (1-809) 431-1600
Tx: WB 2287
C: CARIBANK
Fax: (1-809) 426-7269

Central American Bank for Economic Integration (CABEI)
Apartado Postal 772
Tegucigalpa
Honduras
Phone: (504) 222230
Tx: bancadie 1103
C: BANCADIE

Central American Commission for Environment and Development
(Comision Centroamericana de Ambiente y Desarrollo) (CCAD)
7a Avenida 13-01, Zona 9
Edifico La Cupula
2Q Nivel
Guatemala City
Guatemala
Phone/Fax: (502 2) 343876

Club du Sahel
c/o OECD Development Assistance Committee
2, rue André Pascal
F-75775 Paris Cddex 16
France
Phone: (33 1) 45 24 90 22
Tx: 620160
C: DEVELOPECONOMIE
Fax: (33 1) 45 26 90 31

Colombo Plan Bureau
P.O. Box 596
Colombo 4
Sri Lanka
Phone: (94 1) 581813
Tx: 21537 metalix ce
Fax: (194 1) 580721

Commission for the Conservation of Antarctic Marine Living Resources (CCAMLR)
25 Old Wharf
Hobart, Tasmania 7000
Australia
Phone: (61 02) 310366
Tx: AA57236
C: CCAMLR, HOBART
Fax: (61 02) 232714

Committee of International Development Institutions on the Environment (CIDIE)
P.O. Box 30552
Nairobi, Kenya
Phone: (254 2) 230800
Tx: 22068 unep ke
C: UNITERRA NAIROBI
Fax: (254 2) 226886

Commomwealth of Independent States
(No Secretariat)

Commonwealth Secretariat
Marlborough House, Pall Mall
London SWIY 5HX
England
Phone: (44 71) 839 341 1
Tx: 27678
C:COMSECGEN LONDON SWI
Fax: (44 71)

Conference on Security and Cooperation in Europe (CSCE)
(No permanent Secretariat)

Cooperation Council for the Arab States of the Gulf
(Gulf Cooperation Council) (GCC)
P.O. Box 7153
Riyadh 11462
Saudi Arabia

Phone: (966 1) 4827777
Tx: 403635 tawini sj
Fax: (966 1) 4829089
(Field offices and field stations in Switzerland, Malaysia, Pakistan, Kenya, and Trinidad and Tobago.)

Council of Europe
B.P. 431 R 69 F-67006 Strasbourg Cédex, France
Phone: (33) 88 61 49 61
Tx: eur 870 943
C: EUROPA Strasbourg
Fax: (33) 88 36 70 57

European Community (EC)
Rue de la Loi 200
B-1049 Brussels
Belgium
Phone: (32 2) 235 11
Tx: B 21877
C: COMEUR
Fax: (32 2) 235 01 44

European Investment Bank (EIB)
Boulevard Konrad Adenauer 100
L-2950 Luxembourg-Ville
Luxembourg
Phone: (352) 43791
Tx: 3530
C: BANKEUROP

Federation of Arab Scientific Research Councils (FASRC)
P.O. Box 13027
Baghdad, Iraq
Phone: (964 1) 5381090
C: BAHTHARAB
Tx: 212466 acars ik

Interafrican Committee for Hydraulic Studies (ICHS)
(Comité Interafricain d'Etudes Hydrauliques) (CIEH)
01 B.P. 369
Ouagadougou 01
Burkina Faso
Phone: (226) 30 71 12
Tx: CIEH 5277 BF

Inter-American Development Bank (IDB)
1300 New York Avenue N. W.
Washington, DC
20577
Phone: (202) 623-1000
C: INTAMBANK
Fax: (202) 623-3096

Inter-American Indian Institute (IAII)
Avenida Insurgentes Sur 1690
Colonia Florida
Mexico DF 01030
Phone: (52 5) 660 0007
C: INDIGENI
Fax: (52 5) 534 8090

Inter-American Insitute for Cooperation on Agriculture (IICA)
Apartado 55-2200
Coronado, San Jose
Costa Rica
Phone: (506) 29-0222
Tx: 2144iica
C: IICASANJOSE
Fax: (506) 29-4741

Inter-American Tropical Tuna Commission (IATTC)
c/o Scripps Institution of Oceanography
University of California
La Jolla, California 92093
Phone: (619) 546-7100, ext. 301
Tx: 697 115
Fax: (619) 546-7133

International Association of Fish and Wildlife Agencies
444 North Capitol Street N. W.
Suite 544
Washington, DC 20001
Phone: (202) 624-7890
Fax: (202) 624-7891

International Baltic Sea Fishery Commission (IBSFC)
ul. Hoza 20
PL-00 528 Warsaw
Poland
Phone: (48 22) 28 86 47
Tx: 817421 GOMO PL

International Center for Integrated Mountain Development (ICIMOD)
P.O. Box 3226
Jawalakhel, Lalitpur
Kathmand, Nepal

Phone: (977 1) 525313
Tx: 2439 icimod np
C:ICIMOD, NEPAL
Fax: (977 1) 524509

International Commission for Scientific Exploration of the Mediterranean Sea (ICSEM)
16 Boulevard de Suisse
MC-98000 Monte Carlo Monaco
Phone: (377) 93 30 38 79
Fax: (377) 92 16

International Commission for the Conservation of Atlantic Tuna (ICCAT)
Principe de Vergara 17-7-Q
E-28001 Madrid, Spain
Phone: (34 1) 431 03 29
Tx: 46330 icc CICATUNA
Fax: (34 1) 576 19 68

International Commission for the Southeast Atlantic Fisheries (ICSEAF)
Mali, Morocco, Niger, Pakistan, Qatar, Tunisia, and Turkey

The International Convention for High Seas Fisheries of the North Pacific
65 Paseo de la Habana
E-28036 Madrid, Spain
Phone: (34 1) 458 87 66
Tx: 45533
Fax: (34 1) 571 06

International Council for the Exploration of the Sea (ICES)
(Conseil International pour l'Exploration de la Mer) (CIEM)
Palaegade 2-4
DK-1261 Copenhagen K
Denmark
Phone: +44 (45) 1 15 42 25
Tx: 22498
C: MEREXPLORATION
Fax: +44 (45 1) 93 42 15

International Network of Water Resource Development and Management (INWARDAM)
P.O. Box 815350
Amman, Jordan
Phone: (962 649252
Tx: 23143 iwrn jo

International North Pacific Fisheries Com
6640 Northwest Marine Drive
Vancouver V6TIX2
Canada
Phone:(1-604) 228-1128

International Tropical Timber Organization (ITTO)
8F Sangyo Boeki Centre Building
2 Yamashitacho
Naka-ku, Yokohama 231
Japan
Phone: (81 45) 671 7045
Tx: 3822480 itto j
Fax: (81 45) 671 7007

International Union for Conservation of Nature and Natural Resources (TUCN)
(World Conservation Union)

International Whaling Commission (IWC)
The Red House, Station Road,
Histon, Cambridge CB4 4NP
England
Phone: (44 223) 233971
Fax: (44 223) 232876

Inter-Parliamentary Union (IPU)
C.P. 438
CH-1211 Geneva 19
Switzerland
Phone: (41 22) 734 4150
Tx: 414217
Fax: (41 22) 733 3140

Islamic Development Bank (IDB)
P.O. Box 5925
Jeddah 21432
Saudi Arabia
Phone: (966 2) 6361400
Tx: 601137 isdb sj
C: BANKISLAMI-JEDDAH
Fax: (966 2) 6366871

Latin American and Caribbean Commission on Development and Environment
c/o Inter-American Development Bank
1300 New York Avenue N. W.
Washington, DC 20577

Latin American Housing and Human Settlements Development Organization
(Organizacion Latino Americana de Vivienda y Desarollo de los Sentamientos Humanos) (OLAVI)
(No current information available. Established by the Latin American Economic System, an intergovernmental body of 26 countries headquartered in Caracas, Venezuela.)

League of Arab States (LAS)
37 Avenue Khereddine Pacha
Tunis, Tunisia
Phone: (216 1) 89 01 00
Tx: 13241 tn
Fax: (216 1) 78 18 01

Organization of American States (OAS)
17th Street and Constitution venue N. W.
Washington, DC 20006
Phone: (202) 789-3000

Oslo Commission (OSCOM)
New Court, 48 Carey Street
London WC2A 2JE
England
Phone: (44 71) 242 9927
Tx: 21 185 Bospar G
Fax: (44 71) 831-7427

OECD Development Centre
94 Rue Chardon Lagache
F-75016 Paris
France
Phone: (33 1) 45 24 82 00

Organization of Eastern Caribbean States (OECS)
Natural Resources Management Project (NRMP)
P.O. Box 1383
Castries, St. Lucia
Phone: (1-809) 452-1837
Tx: 6325 oecs nrmplc

OECD Development Center
94 Rue Chardon Lagache
F-75016 Paris
France
Phone: (33 1) 45 24 82 00

Organization for Economic Cooperation and Development (OECD)
2 rue André Pascal
F-75775 Paris
France
Phone: (33 1) 45 24 82 00
Fax: (33 1) 45 24 85 00

Organization of American States (OAS)
17th Street and Constitution Avenue N. W.
Washington, DC 20006
Phone: (202) 789-3000

Pan American Center for Sanitary Engineering and Environmental Sciences
(Centro Panamericano de Ingenieria Sanitaria y Ciencias del Ambiente) (CEPIS)
Casilla Postal 4337
Lima 100, Peru
Phone: (51 14) 354135
Tx: 21052 pe
C: CEPIS
Fax: (51 14) 378289

Pan American Health Organization (PAHO)
(See Organization of American States.)

Paris Commissio (PARCOM)
New Court, 48 Carey Street
London WC2A 2JE
England
Phone: (44 71) 242 9927
Tx: 21185 Bospar G
Fax: (44 71) 831 7427

Permanent Interstate Committee for Drought Control in the Sahel
(Comit6 Permanent Inter-Etats de Lutte contre la S6cheresse dans le Sahel) (CILSS)
B.P. 7049
Ouagadougou
Burkina Faso
Phone: (226 3) 30 67 59
Tx: 5263 comiper ouaga

Permanent South Pacific Commission

(Comision Permanente del Pacifico Sur) (CPPS)
Casilla 16638
Agencia 6400-9
Santiago 9, Chile
Phone: (56 2) 6951 1 01
Tx: 240982 cpps v cl
Fax: (56 2) 6951100

Ramsar Convention Bureau
Avenue du Mont-Blanc
CH-1 196 Gland
Switzerland
Phone: (41 22) 364 91 14
Tx: 419624
Fax: (41 22) 364 83 75

Regional Organization for the Protection of the Marine Environment (ROPME)
P.O. Box 26388
13124 Safat, Kuwait
Phone: (965) 5312140
Tx: 44591 ropme
C: ROPME-KUWAIT
Fax: (965) 5324172

River, Lake, and Boundary Commissions

Africa

Gambia River Basin Development Organization
B. P. 2353
Dakar, Senegal

Lake Chad Basin Commission
B. P. 727
Ndjamena, Chad

Niger Basin Authority
B. P. 729
Niamey, Niger

Organization for the Management and Development of the Kagera River Basin
B. P. 297
Kigali, Rwanda

Permanent Joint Technical Commission for Nile Waters
P.O. Box 878
Khartoum, Sudan

Senegal River Development Organization
46 Rue Carnot
Dakar, Senegal

Latin America

Intergovernmental Committee on the River Plate Basin
Paraguay 755, 2Q Piso
1057 Buenos Aires CF
Argentina

North America

International Boundary and Water Commission
United States and Mexico
c/o U.S. Section
4171 N. Mesa Street
Suite C310
El Paso, TX 79902

International Joint Commission on Boundary Waters
United States and Canada
c/o Canadian Section
Berger Building, 18th Floor
100 Metcalfe Street
Ottawa, Ontario KIP 5MI
Canada

East Asia

Committee for Coordination of Investigations of the Lower Mekong Basin
Pilbultham Villa
Kasatsuk Bridge
Bangkok 10330
Thailand

Europe

Central Commission for the Naviagation of the Rhine
Palais du Rhin
Place de la Republique
F-67000 Strasbourg
France

Danube Commission
Benczur Utca
H-1068 Budapest
Hungary

International Commission for the Protection of Lake Constance
c/o Dipl. Ing. B. Engstle
Bayer Landesamt ffir Wasserwirtschaft

Lazarettstrasse 67
D-8000 Munich 67
Germany

International Commission for the Protection of the Moselle Against Pollution
c/o Ministre des Affaires Etrangeres
Quai d'Orsay
F-75 Paris
France

International Commission for the Protection of the Rhine Against Pollution
Postfach 309
D-5400 Koblentz
Germany

International Commission for the Protection of the Saar Against Pollution
Bundesministerium I Umwelt
Postfach 120629
D-5300 Bonn 1
Germany

International Commission for the Protection of the Waters of Lake Leman Against Pollution
C. P. 80
CH-1000 Lausanne 12
Switzerland

Joint Danube Fishery Commission
Safarikovd 20
CS-01 180 Zilina
Czechoslovakia

Secretariat for the Protection of the Mediterranean Sea
Plaga Lesseps 1
E-08023 Barcelona
Spain
Phone: (34 3) 217 16 95
Tx: 54 519 laye

Southern African Development Coordination Conference (SADCC)
Private Bag 0095, Gaborone
Botswana
Phone: (267 31) 51863

South Asia Cooperative Environment Programme (SACEP)
P.O. Box 1070, Colombo 4
Sri Lanka
Phone: (94 1) 582553
Tx: 21494.global ce
C: SACEP Colombo

South Asian Association for Regional Cooperation (SAARC)
P.O. Box 4222, Kathmandu
Nepal
Phone: (977 1) 221785
Tx: 2561 saarc np
Fax: (977 1) 227033

South Pacific Commission (SPC)
B.P. D5, Noumea CEDEX
New Caledonia
Phone: (687) 26 20 00
Tx: 3139 nm sopacom
C: SOUTHPACOM
Fax: (687) 26 38 18

South Pacific Forum (SPF)
G.P.O. Box 856, Suva
Fiji
Phone: (679) 312 600
Tx: specsuva fj 2229
C: SPECSUVA FJ
Fax: (679) 302 204

South Pacific Regional Environment Program (SPREP)
P.O. Box 240, Apia
Wester Samoa
Phone: (685) 21929
Fax: (685) 20231

Training Center on Environmental Matters for Small Local Authorities in the EEC Mediterranean Countries
(Centre de Formation a l'Environnement pour les Elus et Techniciens Locaux des Pays Mediterrandens de la CEE)
(Defunct organization.)

Tropical Agricultural Research and Training Center
(Centro Agronomico Tropical de Investigacion y Ensefianza) (CATIE)
7170 Turrialba
Costa Rica
Phone: (506) 56-6431

Tx: 8005 catie cr
C: CATIE, Tutrialba
Fax: (506) 56-1533

United States-Panama Joint Commission on the Environment of the Panama Canal
c/o Desk Officer for Panama
Department of State, Room 4906
2201 C Street N. W.
Washington, DC 20520 USA

University for Peace (UP)
.Apartado 138
6100 Ciudad Colon
Costa Rica
Phone: (506) 491324
Fax: (506) 491929

World Tourism Organization (WTO)
Capitdn Haya 42
E-28020 Madrid
Spain
Phone: (34 1) 571 06 28
Tx: 42188 omte
C: OMTOUR Madrid
Fax: (34 1) 57137233

APPENDIX IV

U. S. GOVERNMENT AGENCIES ENGAGED IN INTERNATIONAL PROJECTS AND PROVIDING FOREIGN ASSISTANCE

CABINET DEPARTMENTS

Department of Agriculture

USDA Building
Independence Avenue between 12th and 14th Streets S. W., 20250
General locator phone (202) 720-2791
(All offices are located at the USDA Building unless otherwise indicated.)

Under Secretary for Natural Resources and Environment
Room 217 E
Administration Building
Phone (202) 720-7173

Department of Commerce

Herbert C. Hoover Building
14th Street and Constitution Avenue, 20230
General locator phone (202) 482-2000
(All offices are located at the Herbert C. Hoover Building unless otherwise indicated.)

Under Secretary for International Trade
Room 3850, Phone (202) 482-2867

Assistant Secretary for Export Administration
Room 3886 C, Phone (202) 482-5491

Assistant Secretary for International Economic Policy
Room 3868A, Phone (202) 482-3022

Assistant Secretary for Trade Development
Room 3832, Phone (202) 482-1581

Department of Defense

Assistant Secretary for International Security Policy
Room 4E817, Phone (703) 695-0942
Pentagon, 2600 Defense, Washington DC 26301-2600

Assistant Secretary for Manpower and Reserve Affairs, Installations, and Environment
Room 2E520, Phone (703) 697-6631

1500 Defense Pentagon, Washington, DC 20301-1500

Department of Energy

James Forestal Building
1000 Independence Avenue S. W., 20585
Phone (202) 586-5000
(All offices are located at the James Forrestal Building unless otherwise indicated.)

Department of State

State Department Building
2201 C Street N. W., 20520
Phone (202) 647-4000
(All offices are located at the State Department Building unless otherwise indicated.)

Under Secretary for Internal Security Affairs Room 7208, Phone (202) 647-1049

Assistant Secretary for African Affairs
Room 6234A, Phone (202) 647-4440

Assistant Secretary for East Asian and Pacific Affairs
Room 6206, Phone (202) 647-9596

Assistant Secretary for European and Canadian Affairs
Room 6226, Phone (202) 647-9626

Assistant Secretary for International Organization Affairs
Room 6323, Phone (202) 647-9600

Assistant Secretary for Near Eastern Affairs
Room 6242, Phone (202) 647-9588

Assistant Secretary for South Asian Affairs
Room 6254, Phone (202) 736-4325

INDEPENDENT AGENCIES

African Development Foundation
Suite 1000, 1400 I Street N. W., 20005
Phone (202) 673-3916

Agency for International Development (AID)
320 21st Street N. W.,
Phone (202) 647-1850

American Red Cross
National Headquarters, 17th and D Streets N. W., 20006
Phone (202) 737-8300

Environmental Protection Agency (EPA)
401 M Street S. W., 20460
Phone (202) 260-2090, 800-424-8802 (emergency response)

Export-Import Bank of the United States
Suite 1203, 811 Vermont Avenue N. W.,
Phone (202) 566-8990

Inter-American Foundation
Suite 1000, 901 N. Stuart Street
Arlington, VA 22203
Phone (703) 841-3800

International Bank for Reconstruction and Development (World Bank)
1818 H Street N. W., 20433
Phone (202) 477-1234

International Monetary Fund (IMF)
700 19th Street N. W.,
Phone (202) 623-7759

National Science Foundation
4201 Wilson Boulevard
Arlington, VA 22230
Phone (703) 306-1070

Overseas Private Investment Corporation
1100 New York Avenue N. W., 20527
Phone (202) 336-8400

Peace Corps
1990 K Street N. W., 20526
Phone (202) 606-3886, (800) 424-8590, ext. 293

U.S. Information Agency (USIA)
301 4th Street S. W.,
Phone (202) 619-4355

U.S. International Trade Commission
500 E Street S. W.,
Phone (202) 205-2651

APPENDIX V

FOREIGN EMBASSIES IN THE UNITED STATES

Afghanistan
2341 Wyoming Avenue N. W., 20008
(202) 234-3770

Albania
1150 18th Street N. W., 20006
(202) 223-4942

Algeria
2118 Kalorama Road N. W., 20008
(202) 265-2800

Angola
125 East 73rd Street, New York, N.Y., 10021
(212) 861-5656

Antigua and Barbuda
3007 Tilden Street N. W., 20007
(202) 362-5122

Argentina
1600 New Hampshire Avenue N. W., 20009
(202) 939-6400

Armenia
1660 L Street N. W., Suite 210, 20036
(202) 628-5766

Australia
1601 Massachusetts Avenue N. W., 20036
(202) 797-3000

Austria
3524 International Court N. W., 20008
(202) 895-6767

Azerbaijan
927 15th Street N. W., Suite 700, 20005
(202) 842-0001

Bahamas
2220 Massachusetts Avenue N. W., 20008
(202) 319-2660

Bahrain
3502 International Drive N. W., 20008
(202) 342-0741

Bangladesh
2201 Wisconsin Avenue N. W., Suite 300, 20007
(202) 342-8372

Barbados
2144 Wyoming Avenue N. W., 20008
(202) 939-9200

Belarus
1511 K Street N. W., 20001
(202) 986-1606

Belgium
3330 Garfield Street N. W., 20008
(202) 333-6900

Belize
2535 Massachusetts Avenue
N. W., 20008
(202) 363-4505

Benin
2737 Cathedral Avenue N. W.,
20008
(202) 232-6656

Bhutan
2 UN Plaza, 27th Floor, New York,
NY, 10017
(212) 826-1919

Bolivia
3014 Massachusetts Avenue
N. W., 20008
(202) 483-4410

Botswana
3007 Tilden Street N.W., 20007
(202) 244-4990

Brazil
3006 Massachusetts Avenue
N. W., 20008
(202) 745-2700

Brunei
Watergate, Suite 300, 3rd Floor
2600 Virginia Avenue N. W., 20037
(202) 342-0159

Bulgaria
1621 22nd Street N. W., 20008
(202) 387-7969

Burkina Faso
2340 Massachusetts Avenue
N. W., 2008
(202) 332-5577

Burundi
2233 Wisconsin Avenue N. W.,
Suite 212, 20007
(202) 342-2574

Cameroon
2349 Massachusetts Avenue
N. W., 20008
(202) 265-8790

Canada
501 Pennsylvania Avenue
N. W., 20001
(202) 692-1740

Cape Verde
3415 Massachusetts Avenue
N. W., 20007
(202) 965-6820

Central African Republic
1618 22nd Street N. W., 20008
(202) 483-7800

Chad
2002 R Street N. W., 20009
(202) 462-4009

Chile
1732 Massachusetts Avenue
N. W., 20036
(202) 785-1746

China, People's Republic of
2300 Connecticut Avenue
N. W., 20008
(202) 328-2517

Columbia
2118 Leroy Place N. W., 20008
(202) 387-8339

Comoros
336 E. 45th Street, 2nd Floor, New
York, NY, 10017
(212) 972-8010

Congo, People's Republic of
4891 Colorado Avenue, N. W.,
20011
(202) 726-5500

Costa Rica
2114 S Street N. W., 20008
(202) 234-2945

Cote d'Ivoire
2412 Massachusetts Avenue,
N. W., 20008
(202) 797-0300

Croatia
236 Massachusetts Avenue
N. E., 20002
(202) 543-5580

Cyprus
2211 R Street N. W., 20008
(202) 462-5772

Czech Republic
3900 Linnean Avenue N. W., 20008
(202) 363-6315

Delegation of the Commission of
the European Communities

2100 M Street N. W., 7th Floor, 20037
(202) 862-9500

Denmark
3200 Whitehaven Street N. W., 20008
(202) 234-4300

Djibouti
1156 15th Street N. W., Suite 515, 20005
(202) 331-0270

Dominican Republic
1715 22nd Street N. W., 20008
(202) 332-6280

Ecuador
2535 15th Street N. W., 20009
(202) 234-7200

Egypt, Arab Republic of
2310 Decatur Place N. W., 20008
(202) 232-5400

El Salvador
1010 16th Street N. W., 3rd Floor, 20036
(202) 265-9671

Eritrea
910 17th Street N. W., Suite 400, 20005
(202) 26S-3070

Estonia (Consulate General)
630 5th Avenue, Suite 2415, New York, NY, 10111
(212) 247-1450

Ethiopia
2134 Kalorama Road N. W., 20008
(202) 234-2281

Fiji
2233 Wisconsin Avenue N. W., Suite 240, 20007
(202) 337-8320

Finland
3216 New Mexico Avenue N. W., 20016
(202) 363-2430

France
4101 Reservoir Road N. W., 20007
(202) 944-6000

Gabon
2034 20th Street N. W., 20009
(202) 797-1000

Gambia, The
1155 15th Street N. W., Suite 1000, 20005
(202) 785-1399

Germany
4645 Reservoir Road N. W. 20007
(202) 298-4000

Ghana
3512 International Drive N. W. 20008
(202) 686-4520

Greece
2221 Massachusetts Avenue N. W., 20008
(202) 939-5800

Grenada
1701 New Hampshire Avenue N. W., 20009
(202) 265-2561

Guatemala
2220 R Street N. W., 20008
(202) 745-4952

Guinea
2112 Leroy Place N. W., 20008
(202) 483-9420

Guinea-Bissau, Republic of
918 16th Street N. W., Mezzanine Suite, 20006
(202) 872-4222

Guyana
2490 Tracy Place N. W., 20008
(202) 265-6900

Haiti
2311 Massachusetts Avenue N. W., 20008
(202) 332-4090

Honduras
3007 Tilden Street N. W., 20008
(202) 966-7702

Hungary
3910 Shoemaker Street N. W., 20008
(202) 362-6730

Iceland
2022 Connecticut Avenue N. W., 20008
(202) 265-6653

India
2536 Massachusetts Avenue
N. W., 20009
(202) 939-7000

Indonesia
2020 Massachusetts Avenue
N. W., 20036
(202) 775-5200

Iran
2209 Wisconsin Avenue N. W., 2nd
Floor, 20007
(202) 965-4990

Ireland
2234 Massachusetts Avenue
N. W., 20008
(202) 462-3939

Israel
3514 International Drive N. W.,
20008
(202) 364-5527

Italy
1601 Fuller Street N. W., 20009
(202) 328-5500

Jamaica
850 K Street N. W., Suite 355,
20006
(202) 452-0660

Japan
2520 Massachusetts Avenue
N. W., 20008
(202) 939-6700

Jordan
3504 International Drive N. W.,
20008
(202) 966-2664

Kazakhstan
3421 Massachusetts Avenue
N. W., 20007
(202) 333-4507

Kenya
2249 R Street N. W., 20008
(202) 387-4101

Korea
2600 Virginia Avenue N. W., 2nd
Floor, 20037
(202) 939-5660

Kuwait
2940 Tilden Street N. W., 20008
(202) 966-0702

Kyrgyz Republic
1511 K Street N. W., Suite 705,
20005
(202) 628-0433

Lao People's Democratic Republic
2222 S Street N. W., 20008
(202) 332-6416

Latvia
4325 17th Street N. W., 20011
(202) 726-8213

Lebanon
2560 28th Street N. W., 20008
(202) 939-4300

Lesotho
2511 Massachusetts Avenue
N. W., 20008
(202) 797-5533

Liberia
5303 Colorado Avenue N. W.,
20011
(202) 723-0437

Lithuania
2622 16th Street N. W., 20009
(202) 234-5860

Luxembourg
2200 Massachusetts Avenue
N. W., 20008
(202) 265-4171

Madagascar
2374 Massachusetts Avenue
N. W., 20008
(202) 265-5525

Malawi
2408 Massachusetts Avenue
N. W., 20008
(202) 797-1007

Malaysia
2401 Massachusetts Avenue
N. W., 20008
(202) 328-2700

Mali
2130 R Street N. W., 20008
(202) 332-2249

Malta
2017 Connecticut Avenue
N. W., 20008
(202) 462-3611

Marshall Islands

2433 Massachusetts Avenue
N. W., 20008
(202) 234-5414

Mauritania
2129 Leroy Place N. W., 20008
(202) 232-5700

Mauritius
4301 Connecticut Avenue
N. W., Suite 441, 20008
(202) 244-1491

Mexico
1911 Pennsylvania Avenue
N. W., 20009
(202) 728-1600

Micronesia
1725 N. Street N. W., 20036
(202) 223-4383

Mongolia
2833 M Street N. W., 20007
(202) 298-7137

Morocco
1601 21st Street N. W., 20009
(202) 462-7979

Mozambique
1990 M Street N. W., Suite 570,
20036
(202) 293-7146

Myanmar
2300 S Street N. W., 20008
(202) 332-9044

Namibia
1605 New Hampshire Avenue N.
W., 20009
(202) 986-0540

Nepal
213 1 Leroy Place N. W., 20008
(202) 667-4550

Netherlands
4200 Linnean Avenue N. W., 20008
(202) 244-5300

New Zealand
37 Observatory Circle N. W., 20008
(202) 328-4800

Nicaragua
1627 New Hampshire Avenue N.
W., 20009
(202) 939-6570

Niger
2204 R Street N. W., 20008
(202) 483-4224

Nigeria
2201 M Street N. W., 20037
(202) 822-1500

Norway
2720 34th Street N. W., 20009
(202) 333-6000

Oman
2342 Massachusetts Avenue
N. W., 20008
(202) 387-1980

Pakistan
2315 Massachusetts Avenue
N. W., 20008
(202) 939-6200

Panama
2862 McGill Terrace N. W., 20008
(202) 482-1407

Papua New Guinea
1615 New Hampshire Avenue N.
W., 3rd Floor, 20009
(202) 745-3680

Paraguay
2400 Massachusetts Avenue
N. W., 20008
(202) 483-6960

Peru
1700 Massachusetts Avenue
N. W., 20036
(202) 833-9860

Philippines
1617 Massachusetts Avenue
N. W., 20036
(202) 483-1414

Poland
2224 Wyoming Avenue N. W.,
20008
(202) 234-3800

Portugal 2125 Kalorama Road N.
W. 20008
(202) 332-3007

Qatar
600 New Hampshire Avenue, Suite
1180, 20037
(202) 338-0111

Romania
1607 23rd Street N. W., 20008
(202) 232-4747

Russia
1125 16th Street N. W., 20008
(202) 628-7551

Rwanda
668 New Hampshire Avenue N. W., 20009
(202) 232-2882

Saint Kitts-Nevis
2100 M Street N. W., Suite 608, Washington, DC 20037
(202) 833-3550

Saint Vincent and the Grenadines
1717 Massachusetts Avenue N. W., Suite 102, 20036
(202) 462-7806

Sao Tome and Principe
122 East 42nd Street, Suite 1604, New York, NY, 10168
(212) 697-4211

Saudi Arabia
601 New Hampshire Avenue N. W., 20037
(202) 342-3800

Senegal
2112 Wyoming Avenue N. W., 20009
(202) 234-0540

Seychelles
820 Second Avenue, Suite 900 F, New York, NY, 10017
(212) 687-9766

Sierra Leone
1701 19th Street N. W., 20009
(202) 939-9261

Singapore
1824 R Street N. W., 20009
(202) 667-7555

Slovak Republic
3900 Linnean Avenue N. W., 20008
(202) 363-6315

Solomon Islands
820 Second Avenue, Suite 800, New York, NY, 10017
(212) 599-6193

South Africa
3201 New Mexico Avenue N. W., Suite 300, 20016
(202) 232-4400

Spain
2700 15th Street N. W., 20009
(202) 265-0190

Sri Lanka, Republic of
2148 Wyoming Avenue N. W., 20008
(202) 483-4025

Sudan
2210 Massachusetts Avenue N. W., 20008
(202) 338-8565

Suriname
4301 Connecticut Avenue N. W., Suite 108, 20008
(202) 244-7488

Swaziland
3400 International Drive N. W., 20009
(202) 362-6683

Sweden
600 New Hampshire Avenue N. W., Suite 1200, 20037
(202) 944-5600

Switzerland
2900 Cathedral Avenue N. W., 20008
(202) 745-7900

Syrian Arab Republic
2215 Wyoming Avenue N. W., 20008
(202) 232-6313

Tanzania
2139 R Street N. W., 20008
(202) 939-6125

Thailand
2300 Kalorama Road N. W., 20008
(202) 234-5052

Togo
2208 Massachusetts Avenue N. W., 20008
(202) 234-4212

Trinidad and Tobago
1708 Massachusetts Avenue N. W., 20036

(202) 467-6490

Tunisia
1515 Massachusetts Avenue N. W., 20008
(202) 862-1850

Turkey
1714 Massachusetts Avenue N. W., 20036
(202) 659-8200

Uganda
5909 16th Street N. W., 20011
(202) 726-7100

Ukraine
3350 M Street N. W., 20007
(202) 333-0606

United Arab Emirates
3000 K Street N. W., Room 600, 20007
(202) 338-6500

United Kingdom
3100 Massachusetts Avenue N. W., 20008
(202) 462-1340

Uruguay
1918 F Street N. W., 20006
(202) 331-1313

Venezuela
1099 30th Street N. W., 20007
(202) 342-2214

Western Samoa
1155 15th Street N. W., Suite 510, 20005
(202) 833-1743

Yemen, Republic of
2600 Virginia Avenue N. W., 7th Floor, 20037
(202) 965-4760

Yugoslavia
2410 California Street N. W., 20008
(202) 462-6566

Zaire
1800 New Hampshire Avenue N. W., 20009
(202) 234-7960

Zambia
2419 Massachusetts Avenue N. W., 20008
(202) 265-9717

Zimbabwe
1608 New Hampshire Avenue N. W., 20009
(202) 332-7100

APPENDIX VI
INTERNATIONAL NONPROFIT/ NONGOVERNMENTAL ORGANIZATIONS

African Environmental Research and Consulting Group (AERCG)
14912 Walmer Street
Overland park, Kansas 66203
Phone: (913) 897-6132/ (816) 861-6186
Fax: (913) 897-6132/ (816) 861-0939

Advisory Committee on Pollution of the Sea (ACOPS)
57 Duke Street
Grosvenor Square
London WIM 5DH, England
Phone: (44 71) 493 3092
Founded in 1952 at the initiative of Sir James Callaghan, former British prime minister. ACOPS' goal is to promote "the preservation of the seas of the world from pollution by human activities" through research, publicity, education, and in governmental forums. Provides consultancy services to intergovernmental bodies and NGOS

African Center for Technology Studies (ACTS)
P.O. Box 45917
Nairobi, Kenya
Phone: (254 2) 744047
Fax: (254 2) 743995
Founded in 1988. ACTS' objective is to conduct policy research, undertake training, and disseminate information on science, technology, and environment, with emphasis on sustainable development and the findings of the World Commission on Environment and Development (the Brundtland Report).

African NGOs Environment Network (ANEN)
(Réseau des ONG Africaines sur l'Environnement)
P.O. Box 43844
Nairobi, Kenya
Phone: (254 2) 228138
Tx: 25331 anen ke
Fax: (254 2) 335108
Founded in 1982. ANEN's objective is to strengthen the capacity and technical competence of African

NGOs on environment and development issues.

African Soil Science Association
c/o Department of Soil Science
Faculty of Agriculture
Makerere University
P.O. Box 7062
Kampala, Uganda
Founded in 1986. This organization focuses on information exchange on African soils and their management.

Aga Khan Foundation
Rue Versonnex 7
CH-1207 Geneva, Switzerland
Phone: (41 22) 736-0344
Fax: (41 22) 736-0948
Founded in 1967; part of a network of social development institutions started in the 1890s and chaired by His Highness the Aga Khan, the Iman of the Ismali Muslims.

Alliance of Northern People for Environment and Development (ANPED)
WEED Service Centre
Siegfried-Leopold Stasse 53
D-5300 Bonn, Germany
Phone: (49 228) 470806
Fax: (49 228) 473682
A network organization of NGOs in Europe and North America.

Antarctic and Southern Ocean Coalition
707 D Street S. E.
Washington, DC 20003
Phone: (202) 544-0236
Tx: 158215480
Fax: (202) 543-4710
Founded in 1978. Objective is to protect the Antarctic region by keeping a close watch on governmental and nongovernmental actions regarding the region.

Anti-Slavery International for the Protection of Human Rights
180 Brixton Road
London SW9 6AT
England
Phone: (44 71) 582 40 40
Fax: (44 71) 587-0573
Founded in 1839. A human rights organization with the objective of advancing the well-being of indigenous peoples, as well as the eradication of slavery.

Aquatic Ecosystem Health and Management Society
c/o Cagada Centre for Inland Waters
P.O. Box 5050
Burlington, Ontario L7R 4A6
Canada
Phone: (416) 336-4867
Fax: (I 416) 336-4819
The objective of this organization is to promote and undertake scientific initiatives to ensure the protection of global aquatic environments.

Arctic Institute of North America
University of Calgary
2500 University Drive N. W.
Calgary, Alberta T2N IN4
Canada
Phone: (403) 220-7515
Founded in 1945. Objective of the organization is to provide advocacy, education, and information on the polar regions.

ASEAN Association for Planning and Housing (AAPH)
Strata 100 Building, Ground Floor
Emerald Avenue, 1600 Pasig
Metro Manila
Philippines
Phone: (63 2) 631-3462
Founded in 1979. Objective is to promote the study and practice of regional, urban, and rural planning to improve human settlements and the environment.

Asia Society for Environmental Protection (ASEP)
c/o CDG-South East Asia Program Office (SEAPO)
Asian Institute of Technology
G.P.O. Box 2754, Bangkok 10501
Thailand

Asian Ecological Society

c/o Dr. Jun-Yi Lin
Biology Department
Tunghai University
P.O. Box 843, Taichung 40704
Taiwan
Phone: (886 4) 359-0249
Fax: (886 4) 3595622
Founded in 1977. Goal is to mobilize and encourage Asian scientists to study the ecosystems of their respective nations.

Asian Environmental Society (AES)
U- 1 12 A, Third Floor
Vidhata House, Vikas Marg
Shakarpur, Delhi 110092
India
Phone: (91 11) 224-3500
Founded in 1972. The Society provides a forum for discussion of environmental problems in Asia.

Asian Fisheries Society (AFS)
MC P.O. Box 1501
Makati, Metro Manila 1299
Philippines
Phone: (63 2) 818-0466
Tx: 64794 iclarm pn
Fax: (63 2) 816-3183
Founded in 1984. AFS' objective is to promote interaction and cooperation among fisheries scientists and technologists in Asia.

Asian Forum of Environmental Journalists (AFEJ)
P.O. Box 3094
Kathmandu, Nepal
Phone: (977 1) 470419
Fax: (977 1) 226602
Founded in 1988. Objective is to promote communication between journalists of the region and the Asian countries regarding environmental activities and environmental protection in the region.

Asian Institute of Technology (AIT)
P.O. Box 2754
Bangkok 10501
Thailand
Phone: (66 2) 5160110-29
Tx: 84276th
C: AITBANGKOK
Fax: (66 2) 5162126
AIT promotes the Interdisciplinary Natural Resources Development and Management Program and education on the management of natural resources in the region.

Asian Wetlands Bureau
Institute of Advanced Studies
University of Malaya
Lembah Pantai
59100 Kuala Lumpur
Malaysia
Phone: (60 3) 7572176
Tx: unimal ma 39845
Fax: (60 3) 7573661
This organization promotes sustainable use and conservation of wetland resources within the Asian region.

Asia-Pacific People's Environment Network (APPEN)
c/o Sahabat Alam Malaysia
43 Salween Road
10050 Penang, Malaysia
Phone: (60 4) 376930
Tx: cappg ma 40989
Founded in 1983. The network's objective is to organize diverse NGOs and individuals in the Asian Pacific region to educate and to collect and disseminate information on environmental issues.

Association Internationale Futuribles
(International Futures Research Association)
55 Rue de Varenne
F-75341 Paris Cddex 07
France
Phone: (33 1) 42 22 63 10
Tx: 201220-F
Fax: (33 1) 42 22 65 54
Promotes and conducts future studies on new technologies, social policies, aging population, lifestyle changes; provides information on society.

Association of Baltic National Parks
(Balti Rahvusparkide Assotsiatsioon)

Lahemaa Rahvuspark
EE-2128 Viitna
Laane-Virumaa, Estonia
Phone: (7 014 32) 4 57 59
Fax: (7 014 2) 45 33 10 "for Lashemaa".

Association of Geoscientists for International Development (AGID)
c/o Asian Institute of Technology
P.O. Box 2754
Bangkok 10501
Thailand
Phone: (66 2) 5160110-44, ext. 5514
Tx: 84276 th.
C: AIT-BANGKOK TH
Fax: (66 2) 5162126
Founded in 1974. AGID promotes activities in the geosciences and management of natural resources related to the needs of developing countries.

Athens Center of Ekistics (ACE)
24 Stratiotikou Syndesmou St.
P.O. Box 3471
102-10 Athens, Greece
Phone: (30 1) 3623216
Tx: 215227
C: ATINST
Fax: (30 1) 3633395
Founded in 1963. Promotes research, education, documentation, and international cooperation in the study of all aspects of human settlements.

Atlantic Salmon Federation (ASF)
P.O. Box 429
St. Andrews, New Bruswick EOG 2XO
Canada
Phone: (506) 529-8889
Fax: (506) 529-4438
Founded in 1982. Promotes the preservation, conservation, education for international cooperation, and the management of the Atlantic salmon and its habitat.

Bahiti International Community
Office of the Environment
866 United Nations Plaza
Suite 120, New York, NY 10017
Phone: (212) 486-0560
Fax: (212) 838-7027
Advocates for the environmental interests and concerns of the Bahd'i faith.

Bank Information Center
2025 1 Street N. W., Suite 522
Washington, DC 20006
Phone: (202) 466-8191)
Fax: (202) 466-8189
Founded in 1987. Provides information and a clearinghouse for NGOs regarding the operations of the World Bank and other multilateral development banks and their impact on the environments of developing countries.

Bat Conservation International (BCI)
P.O. Box 16203
Austin, TX 78716
Phone: (512) 327-9721
Fax: (512) 327-9724
Founded in 1982. Promotes conservation and protection of bats, as well as education on bats, worldwide.

The Biodiversity Coalition
P.O. Cygnet
Tasmania 7112
Australia
Phone: (61 2) 952745
Fax: (61 2) 951964
Founded in 1992. An international NGO network working for biodiversity conservation.

Buddhist Perception of Nature
5H Bowen Road, 1st Floor
Hong Kong
Phone: (852 5) 233464
Tx: 72149 sidan hx
Fax: (852) 691619
Founded in 1985. Conducts research on and assembles "traditional Buddhist teachings regarding man's interdepen-

dence with and/or responsibilities to the natural environment and all living things."

Business Council for Sustainable Development
World Trade Centre, 10
Route de l'Atroport, CH1215 Geneva 15
Switzerland
Phone: (41 22) 788 3202
Fax: (41 22) 788 321 1
Formed in 1991.

Canada-United States Environmental Council
1244 19th Street N. W.,
Washington, DC 20036
Phone: (202) 659-9510
Founded in 1971. A group of major Canadian and U.S. NGOs.

Caribbean Conservation Association
Savannah Lodge, The Garrison
St. Michael, Barbados
Phone: (809) 426-5373
Fax: (809) 429-8483
Founded in 1967. Promotes conservation and development of the environment and preservation of the cultural heritage of the Caribbean region.

Caribbean Conservation Corporation (CCC)
P.O. Box 2866
Gainesville, FL 32602
Phone: (904) 373-6441
Fax: (904) 375-2449
Founded in 1959. Operates a research station in Tortuguero, Costa Rica, and a seminatural impoundment for turtles in the Bahamas.

Caribbean Natural Resources Institute (CANARI)
1104 Strand Street, Suite 206
Christiansted, St. Croix
U.S. Virgin Islands 00820
Fax: (809) 773-9854
Clarke Street, Vieux Fort, St. Lucia
Phone: (809) 454-6060
Fax: (809) 454-5188
Founded in 1977.

International Primate Protection League (IPPL)
P.O. Box 766
Summerville, SC 29484
Phone: (803) 871-2280
Fax: (803) 871-7988
Founded in 1973. Promotes education and protection of primates such as apes, lemurs, monkeys.

International Primatological Society (IPS)
c/o Deutsches Primatenzentrum
Kellnerweg 4
D-3400 Göttingen
Germany
Founded in 1964. IPS' objectives are to conduct research and ensure the protection of primates.

International Professional Association for Environmental Affairs (IPRE)
31, Rue Montoyer, Boite 1
B-1040 Brussels
Belgium
Phone: (32 2) 513 60 83
Fax: (32 2) 514 33 86
Founded in 1976. Objective is to provide communication of environmental issues to the public.

International Radiation Protection Association (IRPA)
c/o C. J. Liuyskens, Executive Officer, IRPA
P.O. Box 662
5600 AR Eindhoven
Netherlands
Phone: (31 40) 473355
Tx: 51163 thehv nl
Fax: (31 40) 435020
Founded in 1966. Promotes the protection of human health and the environment from the hazards of ionizing radiation.

International Reference Center for Community Water Supply and Sanitation

International Research Center on the Environment Pio Manzu
Pio Manzu International Research Center

International Rivers Network (IRN)
1847 Berkeley Way
Berkeley, CA 94703
Phone: (510) 848-1155
Fax: (510) 848-1008
Objective is to protect freshwater resources, endangered ecosystems, and the rights of indigenous peoples worldwide.

International Social Science Council (ISSC)
1 Rue Miollis
F-75015 Paris, France
Phone: (33 1) 45 68 25 58
Tx: 204 462 unesco Paris, F
Fax: (33 1 43 06 87 98
Founded in 1952. Objectives are to study the social dimensions of resource use and to assess global environmental conditions.

International Social Sciences and Environment Network
(Reseau International Sciences Sociales et Environnement) (RISE)
c/o Dominique Allan-Michaud, Institut de Politiques Internationale et Européene
31 Rue de Chazelles F-75017
Paris, France
Phone: (33 1) 46 22 40 23
Founded in 1984. Objective is to promote environmental development.

International Society for Ecological Economics (ISEE)
P.O. Box 1589
Solomons, MD 20688
Phone: (410) 326-0794
Fax: (410) 326-6342
Objective is to promote the integration of economic and environmental policies.

International Society for Ecological Modeling (ISEM)
International Secretariat
DK3500 Vmerlose
Copenhagen, Denmark
Phone: (45) 42 48 06 00
Fax: (45) 35 37 57 44
Founded in 1975. Promotes ecological and environmental modeling through computer technology.

International Society for Endangered Cats (ISEC)
4638 Wintersdt Drive
Columbus, OH 43220
Phone: (614) 451-4460
Fax: (614) 451-9221
Founded in 1988. Objective is to aid the preservation and conservation of endangered and threatened cat species through research and education.

International Society for Environmental Education
c/o School of Planning and Architecture
4 Block B
Indraprashta Estate
New Delhi 110002, India
Founded in 1972.

International Society for Environmental Ethics
c/o Prof. Laura Westra, Secretary
Department of Philosophy
University of Windsor
Windsor, Ontario N9B 3P4
Canada
Phone: (519) 253 4232, ext. 2342
Fax: (519) 973-7050
An organization of environmental consultants, lawyers, policymakers, and other professionals, and students.

International Society for Environmental Management
c/o Bundesdeutscher Arbeitskreis Für Umweltbewusstes Management (BAUM)
(German Environmental Management Association)
Sillemstrasse 36
D-2000 Hamburg 20

Phone: (49 40) 4911814
Objective is to link various national associations for environmental management.

International Society for the Prevention of Water Pollution
Little Orchard, Bentworth
Alton, Hampshire GU34 5RB
Phone: (44 252) 310113 (A.M.), 624837 (P.M.)
Investigates water pollution worldwide and suggests solutions; operates through a network of members and correspondents.

International Society for Research on Civilization Diseases and Environment
(Société Internationale pour la Recherche sur les Maladies de Civilisation et l'Environnement)
(SIRMCE) B.P. 17159
L-1017 Luxembourg
Phone: (352) 49 53 70
Founded in 1973. Objective is to contribute to research on civilization diseases that threaten the health of humanity.

International Society for Tropical Ecology (ISTE)
c/o Department of Botany
Banaras Hindu University
Varanasi 221005 UP
India
Phone: 5429 1, ext. 352
Founded in 1960. Promotes the study of ecology and conservation of natural resources in the tropics and subtropics.

International Society of Biometeorology (ISB)
Witikonerstrasse 440
CH-8053 Zurich, Switzerland
Phone: (43 1) 381 45 59.
Founded in 1956. An organization of biometeorologists who specialize in the relation between climate and life, working in the fields of agriculture; botany; forestry; entomology; and human, veterinary, and zoological biometeorology.

International Society of City and Regional Planners (ISOCARP)
Mauritskade 23
N-2514 HD The Hague
Netherlands
Phone: (31 70) 346 26 54
Fax: (31 70) 361 79 09
Founded in 1965. Promotes the profession of city and regional planning and fosters education and research in the field.

International Society of Soil Science (ISSS).
Institute Bodenforschung und Bauge6logiei
Universitdt ffir Bodenkulture
Gregor Mendelstrasse 33
A-1180 Vienna, Austria
Phone: (43 1) 3106026
Tx: via 111010 tzsta
C: ISSSAUSTRIA Wien
Fax: (43 1) 3106027
Founded in 1924. Promotes the scientific study of soils.

International Society of Tropical Foresters (ISTF)
5400 Grosvenor Lane
Bethesda, MD 20814
Phone: (301) 897-8720
Founded in 1950.

International Soil Reference and Information Centre (ISRIC)
P.O. Box 353
6700 AJ Wageningen
Netherlands
Phone: (31 8370) 19063
Tx: 45888 (via IAC)
C: ISOMUS
Fax: (31 8370) 24460
Founded in 1966. Provides technical assistance in soil mapping and classification.

International Solid Wastes and Public Cleansing Association (ISWA)
Vester Farimagsgade 29
DK-1780 Copenhagen V

Denmark
Phone: (45 33) 15 65 65
Fax: (45 1) 93 71 71
Promotes the adoption of acceptable systems of solid wastes management and of public cleansing, through technological development, protection of the environment and conservation of materials and energy resources.

International Studies Association (ISA)
David M. Kennedy Center for International Studies
216 HRCB, Brigham Young University
Provo, Utah 84602
Phone: (801) 378-5459
Tx: 706448 byu
Fax: (801) 378-7075
Founded in 1959. Objective is to facilitate cultural and scientific exchange and channel of communication by providing links between educators, researchers, and practitioners who are interested in the changing global system.

International Training Center for Water
Resources Management (ITCWRM)

International Union for Conservation of Nature and Natural Resources (IUCN)
Rue de Mauverney 28
CH-1196 Gland, Switzerland
Phone: 41 22 999 0001

International Union for the Scientific Study of Population (IUSSP)
Rue des Augustins 34
B-4000 Liege
Belgium
Phone: (32 41) 22 40 80
Tx: 42648
Fax: (32 41) 22 38 47
Founded in 1928. Promotes demography advocacy through international cooperation.

International Union of Air Pollution Prevention Associations
136 North Street
Brighton BNI IRG
England
Phone: (44-273) 26313
C: POLLUTION
Fax: (44-73) 735802
Founded in 1964. Organization promotes public education in regard to global clean air.

International Union of Alpinist Associations.
(Union Internationale des Associations d'Alpinisme) (UIAA)
c/o SAC-Geschdftsstelle
Helvetiaplatz 4
CH3005 Berne, Switzerland
Phone: (41 31) 44 46 24
Tx: 912 388
Founded in 1932. Objective is the protection of mountain environments.

International Union of Anthropological and Ethnological Sciences (IUAES)
University College of North Wales
Bangor, Gwynedd
Wales LL57 2DG
Phone: (44 248) 351151, ext. 482
Tx: 61 1 00 ucnwsl g
IUAES is concerned with anthropology and ethnology worldwide and promotes the involvement of local and indigenous resource management systems and women's participation in environmental movements.

International Union of Architects
(Union Internationale des Architects) (UIA)
51 Rue Raynouard
F-75016 Paris, France
Phone: (33 1) 45 24 36 88)
C: UNIARCH
Founded in 1948. Objective is to strengthen professional ties between architects of all nations, schools, and philosophies, and to promote progressive ideas in the fields of architecture and town planning worldwide.

International Union of Biological Sciences (IBS)
51 Boulevard de Montmorency
F-75016 Paris, France
Phone: (33 1) 45 25 00, 09
Tx: 645554 f
Fax: (33 1) 45 25 20 29
Founded in 1919. Promotes the biological sciences; facilitates and coordinates research and other scientific activities through global cooperation.

International Union of Directors of Zoological Gardens (IUDZG)
c/o First Secretary Bent Jorgensen
Copenhagen Zoo
Sdr. Fasanvej 79
DK-2000 Frederiksberg
Denmark
Phone: (45) 36 30 25 55
Fax: (45) 36 44 24 55
Founded in 1946. Objective is to provide zoo conservation programs.

International Union of Forestry Research Organizations (IUFRO)
A-1131 Vienna
Austria
Phone: (43 1) 82 01 51
Tx: 75312646 iusc a
Fax: (43 1) 82 93 55
Founded in 1892. Promotes international cooperation in scientific studies in forestry.

International Union of Geodesy and Geophysics (IUGG)
c/o G. Balmino, Secretary General, CNES
18 Avenue Edouard Belin
F-31055 Toulouse Cédex, France
Phone: (33 6f 33 28 89
Tx: 531081 f
Fax: (33) 61 25 30 98
Founded in 1919. Promotes scientific study of the earth and applications of the knowledge gained by such studies to the needs of socioenvironmental preservation.

International Union of Pure and Applied Chemistry (IUPAC)
Bank Court Chambers 2-3 Pound Way
Cowley Centre
Oxford OX4 3Y5, England
Phone: (44 865) 747744
Founded in 1919. Promotes and maintains commissions on water chemistry, atmospheric chemistry, and pesticide chemistry.

International Union of Societies of Foresters (IUSF)
c/o Institute of Foresters of Australia
GPO Box 11697
Brisbane, Queensland 4001
Australia
Phone: (61 7) 234-0156
Fax: (61 7) 221-4713
Founded in 1966. Objective is to promote cooperation for advancement of forestry and foresters.

International Union of Speleology
(Union Internationale de Spéléologie)
(UIS)
c/o Dr. C. M. Ek
Rue des Vennes 131
B-4020 Liege, Belgium
Phone: (32 41) 42 00 80
Founded in 1964. Promotes and maintains a protection and management department.

International Water and Sanitation Center
(Centre International de l'Eau et l'Assainissement)
P.O. Box 93190
2509 AD The Hague
Netherlands
Phone: (31 70) 3314133
Tx: 33296 irc nt
C: Worldwater The Hague
Fax: (31 70) 3814034
Founded in 1968. The center's objective is to provide information and documentation services, training, and research and development support for water supply and sanitation.

International Waterfowl and Wetlands Research Bureau (IWRB)
Slimbridge, Glos
GI2 7BX, England
Phone: (44 45) 398 333
Tx: 437145 wwf g
Fax: (44 45) 389 827
Founded in 1954. IWRB collaborates with other international and national organizations to develop and implement activities and programs for the global conservation of wetlands and their waterfowl.

International Water Resources Association (IWRA)
205 N. Matthews
University of Illinois, Urbana, IL 61801
Phone: (217) 333-6275
Fax: (217) 244-6633
Founded in 1972. A global organization for water managers, scientists, planners, manufacturers, administrators, educators, lawyers, physicians, and others concerned with the future of global water resources.

International Water Supply Association (IWSA)
1 Queen Anne's Gate
London SWIH 9BT
England
Phone: (44 71) 222 81 1 1
Fax: (44 71) 222 7243
Founded in 1949. Objective is to improve knowledge and practice in the public supply of water.

International Wilderness Leadership Foundation
(The WILD Foundation)
211 W. Magnolia
Fort Collins, CO 80521
Phone: (303) 498-0303
Tx: 9103506369
Fax: (I 303) 498-0403
Founded in 1974. Objective is to protect wilderness and wildlife.

International Wildlife Coalition (IWC)

International Work Group for Indigenous Affairs (IWGIA)
Fiolstroede 10
DK-1 171 Copenhagen K
Denmark
Phone: (45 33) 12 47 24
Fax: (45 33) 14 77 49
The organization's objective is to support indigenous peoples in their struggle against oppression.

International Youth Conference on the Human Environment (IYCHE)
A periodic conference first held in 1977. No current information available.

International Youth Federation for Environmental Studies and Conservation (IYF)
Astridsvej 7-5-1
DK-8220 Braband
Denmark
Phone: (45 86) 25 01 90
Founded in 1956. Objective is to voluntarily promote youth movement for environmental conservation worldwide.

Inuit Circumpolar Conference (ICC)
650 32nd Avenue, Suite 404
Lachine, Quebec H8T 3K4
Canada
Phone: (514) 637-3771
Fax: (514) 637-3146
Founded in 1977. Objective is to promote long-term management and protection of arctic and subarctic wildlife, environment, and biological productivity.

Islamic Academy of Sciences (IAS)
P.O. Box 830036
Amman, Jordan
Phone: (962 6) 822104
Tx: 24368 ias jo
Founded in 1986. Promotes interaction between Muslim intellectuals in science and technology worldwide.

Latin American Center of Social Ecology
(Centro Latino Americano de Ecologia Social) (CLAES)
Casilla de Correo 13000
11700 Montevideo

Uruguay
Tx: 23391 uy attn CLAES
Fax: (598 2) 921117 attn CLAES
The organization is devoted to the study of human-human and human-environment relationships and the linkages with social and environmental factors.

Latin American Committee on National Parks
(Comité Latinoamericano de Parques Nacionales) (CLAPN)
Florida 165, 4.Q Piso
Oficina 432
1333 Buenos Aires, Argentina
Founded in 1964. Goal is to organize courses, seminars, and technical meetings for personnel of national parks in the Latin American region.

Latin American Consortium on Agroecology and Development
(Consorcio Latinoamericano sobre Agroecologia y Desarrollo) (CLADES)
c/o CET, Casilla 16557
Correo 9
Santiago, Chile
Founded in 1989. Promotes agroecology in rural development through training, cooperative research, and information exchange.

Latin American Federation of Young Environmentalists
(Federacion Latinoamericana de Juvenes Ambientalistas) (FLAJA)
Apartado Adreo 57668
Bogota, Colombia
Founded in 1981. Promotes youth involvement in conservation and optimal use of natural resources.

Latin American Network of Environmental NGOs
(Red Latinoamercana de ONGs Ambientalistas)
c/o CIPFE, Casilla Correo 13125
11700 Montevideo
Uruguay
Fax: (598 2) 985959
Founded in 1985. Objective is to provide clearinghouse and exchange of information and joint activities among Latin American NGOs.

Latin American Network of Social Ecology
(Red Latino Americana de Ecologia Social) (RedLAES)
c/o CLAES, Casilla De Correo 13000
11700 Montevideo
Uruguay
Tx: 23391 uy atn. claes
Fax: (598 2) 92117 atn. CLAES
Founded in 1989. Promotes communication between NGOs, government agencies, and other groups on human-environment relations from a social ecology perspective.

Latin American Plant Sciences Network
(Red Latinoamericana de Botánica) (RLB)
c/o Dra. Mary T. Kalin De Arroyo
Facultad de Ciencias
Universidad de Chile
Casilla 653
Santiago, Chile
Phone/Fax: (56 2) 271-5464
Founded in 1988. Goal is to advance and advocate graduate studies and research in the plant sciences in the Latin American region.

Latin American Social Science Council
Consejo Latinoamericano de Ciencias Sociales
Callao 875, 3er. Piso
1023 Buenos Aires
Argentina
Phone: (54-1) 811-6588
Fax (54-1) 812-8459
Founded in 1967. Objective is to conduct social science research in Latin America.

Latin American Society of Soil Science
(Sociedad Latinoamericana de la Ciencia del Suelo)
c/o Prof. I. Pla-Sentis

Las Acacias, Apartado 1131
Maracay, Venezuela

Law Association for Asia and the Pacific (LAWASIA)
G.P.O. Box A35, Perth
Western Australia 6001
Australia
Phone: (61 9) 221 2303
Fax: (61 9) 221 5914
Founded in 1966. Promotes the rule of the law through developing the law, legal education, and public information.

Men of the Trees (MotT)
Sandy Lane, Crawley Down, Crawley RHIO 4HS
England
Phone: (44 342) 712536
Fax: (44 342) 718282
Founded in 1922. Objective is to promote large-scale tree planting throughout the world and conservation of existing forests.

Mesoamerican Center for the Study of Appropriate Technology
(Centro Mesoamericano de Estudios sobre Tecnologia Apropiada) (CEMAT)
Apartado Postal 1160
Guatemala, Guatemala
Phone: (502 2) 762355
Founded in 1977. Objective is to promote environmental improvement and conservation of natural resources in Central America and Mexico.

Minority Rights Group (MRG)
379 Brixton Road
London SW9 7DE
England
Phone: (44 71) 978 9498
Founded in 1970. Advocates for human rights and publicizes the violations of the rights of minority and majority groups worldwide.

New World Dialogue on Environment and Development in the Western Hemisphere
Secretariat, World Resources Institute
1709 New York Avenue N. W.
Washington, DC 20006
Phone: (202) 638-6300
Fax: (202) 638-0036
Founded in 1990. Promotes and advocates specific international commitments by the hemisphere's governments in order to secure economic development that is environmentally sustainable and socially equitable.

No More Bhopals Network
No current information available.

Nordic Council for Ecology
c/o Dr. Pehr H. Enckell
Ecology Building
University of Lund
S-223 62 Lund
Sweden
Phone: (46 46) 14 81 88
Founded in 1965. Promotes education and research in the science of ecology in the region.

Nordic Council for Wildlife Research
c/o J. O. Pettersson
Uretarna
S-737 91 Ragersta
Sweden
Founded in 1971. Promotes cooperation among wildlife researchers.

Nordic Forestry Federation
c/o Sveriges Skogsvardsforbund
P.O. Box 500
S182 15 Dandengd
Sweden
Phone: (46 8) 755 86 02
Founded in 1946. Promotes cooperation among Nordic countries in the field of forestry.

Nordic Society for Radiation Protection
c/o National Institute of Radiation Hygiene
P.O. Box 55, N-1345 Osteras
Norway
Phone: (47 2) 24 4190
Fax: (47 2) 24 74 07

Founded in 1964. Provides exchange of information and experience on radiation protection.

North American Association for Environmental Education
1255 23rd Street, NW
Suite 400
Washington, DC 20037-1199
Phone: (202) 884-8913
Fax: (202) 884-8701
Promotes environmental education programs at all levels; coordinates environmental education activities among programs and educational institutions; disseminates information; assists in program development.

North American Coalition on Religion and Ecology (NACRE)
5 Thomas Circle N. W.
Washington, DC 20005
Phone: (202) 468-2591
NACRE's goal is to communicate its "Caring for Creation" vision to people throughout North America and the world

1000 Communes for the European Environment
(1000 Communes pour l'Environnement Européen)
c/o France Nature Environment
57 Rue Cuvier, F-75231 Paris Cédex 05
France
Phone: (33 1) 43 36 79 95
Tx: 260921
Founded in 1987. Promotes communication on local environmental issues among local authorities.

Organization for Flora Neotropica (OFN)
c/o Scott Mori
New York Botanical Garden
Bronx, New York 10458
Phone: (212) 220-8742
Tx: 430880 nyb bot gdns nyk
Fax: (212) 220-6504
Founded in 1964. Promotes cooperation among botanists.

Organization for the Phyto-Taxonomic Investigation of the Mediterranean Area (OPTIMA)
c/o Botanischer Garten und Museum
Königin-Luise-Strasse 6-8
D-1000 Berlin 33
Germany
Phone: (49 30) 83006132
Tx: 183798
Fax: (49 30) 83006218
Founded in 1974. Promotes and conducts research, exploration, resource studies, and conservation related to the plant life of the Mediterranean area.

The Organization for Tropical Studies, Inc. (OTS)
P.O. Box DM, Duke Station
Durham, NC 27706
Phone: (919) 684-5774
Fax: (919) 684-5661
Promotes graduate training and research and conservation of natural resources.

The Other Economic Summit (TOES)
P.O. Box 12003
Austin, TX 78711
Phone: (512) 476-4130
Fax: (512) 476-4759
TOES is an international network of independent but cooperating organizations on economic initiatives.

Oxfam
274 Banbury Road
Oxford OX2 7D2
England
Phone: (44 865) 31131 1
Tx: 83610
Fax: (44 865) 312600
Founded in 1942. A major private supplier of aid to developing countries; two thirds of its funds are spent for long-term development projects.

Pacific Science Association (PSA)
P.O. Box 17801
Honolulu, HI 96817
Phone: (808) 848-4139

Founded in 1920. Provides a multidisciplinary forum for discussing common scientific concerns in the Pacific Basin.

Pan-African Council for Protection of the Environment and for Development
(Conseil Panafricain pour la Protection de l'Environnement et le Development)
c/o Dr. Wa Nsanga
B.P. 994
Nouakchott, Mauritania
Phone: 530 77
C: CPPED, Nouakchott
Tx: NOSOM-LIPAM 656/mtn
Founded in 1982. Objective is to combat desertification with a multidiciplinary approach that incorporates the objectives and needs of local communities.

Panos Institute
9 White Lion Street
London NI 9PD
England
Phone: (44 71) 278 1 1 1 1
Tx: 9419293 panos g
Fax: (44 71) 278 0345
Founded in 1986. Promotes international partnership with others towards an understanding of sustainable development worldwide.

Parliamentarians for Global Action
211E 43rd Street, Suite 1604
New York, NY 10017
Phone: (212) 687-7755
Fax: (212) 687-8409
Promotes sustainable development and disarmament.

People and Physical Environment Research (PAPER)
c/o Ross Thorne
Faculty of Architecture
University of Sydney
Sydney, NSW 2006
Australia
Phone: (61 2) 692 2826
Fax: (61 2) 692 3031
Promotes and conducts systematic study of the relationship between people and their physical environment and the application of such knowledge to design and planning.

Pesticide Action Network (PAN)
965 Missouri Street, Suite 514
San Francisco, CA 94103
Phone: (415) 541-9140
Fax: (415) 541-9253
Founded in 1982. Objective is to advocate against the overuse and misuse of pesticides.

Pio Manzii International Research Centre
(Centro Ricerche Pio Manzii)
1-47040 Verucchio
Italy
Phone: (39 541) 678 139
Tx: 550423 cirsa i. C: PIO MANZU
Fax: (39 541) 670 172
Founded in 1969. Promotes research into the effects of technology and industry on the human and cultural environment.

Regional Environmental Center for Central and Eastern Europe (REC)
Miklōs tér 1
1035 Budapest
Hungary
Phone: (36 1) 168-6284
Fax: (36 1) 168-7851
An independent, nongovernmental organization founded in 1990. Objective is to address the environmental challenges common to Central and Eastern Europe.

Regional Mangrove Information Network
No current information available.

Regional Network on Nongovernmental Conservation Organizations for Sustainable Development in Central America
(Red Regional de Organizaciones Ambientalistas No Gubernamentales para el Desarrollo Sostenible de Centroamerica) (REDES-C.A.)
c/o Lic. Juan Jose Montiel R.
Apartado Postal 2431

Managua, Nicaragua
Phone/Fax: (505 2) 74563
Founded in 1987. Promotes and coordinates activities to heighten the level of interaction and communication among Central American NGO.

Renewable Energy and Environmental Conservation Association in Developing Countries (REECA)
c/o 5 Pynnersmead, Herne Hill
London SE24 9LU
England

Sea Shepherd Conservation Society
1314 Second Street
Santa Monica, CA 90401
Phone: (310) 394-3108
Fax: (310) 394-0360
Founded in 1977. Objective is to conserve and protect wildlife, especially marine mammals.

Seeds Action Network International (SAN)
Founded 1985. An informal network of organizations and individuals active in campaigns and field work on genetic resources and biotechnology. Contact through Genetic Resources Action International, which see.

Genetic Resources Action International (GRAIN)
Girona 25, Principal, 08010
Barcelona, Spain
Phone: (34 3) 301-1381
Fax: (43 3) 30116 27

Social Forestry Network
c/o Agricultural Administration Unit
Overseas Development Institute
Regent's College
Inner Circle, Regent's Park
London NWI 4NS
England
Phone: (44 71) 487 7413
Tx: 94082191 odi uk
Fax: (44 71) 487 7790

Societas Europaea Lepidopterologica (SEL)
(European Lepidopterological Society)
c/o Zoologisches Institut, Universitat Bern
Baltzerstrasse 3
CH-3012 Bern
Switzerland
Founded in 1978. Promotes the study and conservation of butterflies and moths and their habitats.

Society for Conservation Biology
c/o Prof. Stanley Temple
Department of Wildlife Ecology
University of Wisconsin
Madison, WI 53706
Phone: (609) 262-2671
Founded in 1985. Objective is to provide scientific information and expertise to promote the preservation of biological diversity.

Society for Ecological Restoration (SER)
1207 Seminole Highway
Madison, WI 53711
Phone: (608) 262-9547
Fax: (608) 262-5209
Founded in 1987. Objective is to develop strategies and conduct activities to define the mutual and beneficial relationships between human beings and the rest of nature.

Society for Ecology
c/o Prof. Dr. Reinhard Bornkamm
Institut for Okologie
Technische Universittit Berlin
Rothenburgstrasse 12
D-1000 Berlin 41
Germany
Phone: (49 30) 31471355
Fax: (49 30) 31471324
Founded in 1970. Advocates and promotes environmental issues concerned with global ecology.

Society for International Development (SID)
Palazzo Civilta del Lavoro
1-00144 Rome
Italy
T: (39 6) 591 7897
Tx: 616484 sid iI
C: SOCINTDEV
Fax: (39 6) 591 9836

Founded in 1957. Promotes and provides communication between individuals and organizations committed to social and economic development.

Society for the Advancement of Socioeconomics (SASE)
714H Gelman Library
2130 H Street N. W.
Washington, DC 20052
Phone: (202) 994-8167
Founded in 1989. Promotes development of a model of economic behavior that includes social science variables and ethical considerations.

Socio-Ecological Union
ul. Krasnoarmeyskaya 25, kv. 85
125319 Moscow
Phone: (7 905) 151 62 70
Fax: (7 095) 200 22 16
Founded in 1988.

SOS Sahel International
c/o SOS Sahel International France
94 Rue Saint Lazare
F-75009 Paris
France
Phone: (33 1) 42 85 08 44
Tx: 281 174 f
Fax: (33 1) 45 96 07 15
Founded in 1976 in Dakar, Senegal. Objective is to inform public agencies and private donors about the drought in the Sahel region of Africa.

South Center
Geneva office: Case Postale 228
CH-1211 Geneva 19
Switzerland
Phone: (41 22) 798-3433
Tx: 415 616 soc ch
Fax: (41 22) 798-8531
Dar es Salaam office: P.O. Box 71000
Dar es Salaam
Tanzania
Phone: (255 52) 46924
Tx: 41 846
Fax: (255 51) 46146
Founded in 1990. Objective is to develop and promote recommendations on development issues from a third-world perspective.

South Pacific Action Committee on Human Ecology and Environment
c/o Institute of Natural Resources
University of the South Pacific
P.O. Box 1168
Suva, Fiji
Phone: (679) 212465

Special Libraries Association (SLA)
1700 18th Street N. W.
Washington, DC 20009
Phone: (202) 234-4700
Fax: (202) 265-9317
SLA is an international professional association serving the needs of special libraries and information professionals.

Speleological Federation of Latin America and the Caribbean
(Federacion Espeleologica de America Latina y el Caribe) (FEALC)
c/o Sociedad Venezolana de Espeleologia
Apartado 47334
Caracas 1041-A
Venezuela
T/Fax: (58 2) 6627845
Founded in 1981. Promotes exchange of information about, and conservation of, caves and karst areas in the region.

Third World Academy of Sciences (TWAS)
c/o International Centre for Theoretical Physics
P.O. Box 586
1-34100 Trieste
Italy
Phone: (39 40) 224 0328
Tx: 460392 ictp i
C: CENTRATOM
Fax: (39 40) 224 559
Founded in 1983. Works to aid in the development of and to support high-level scientific talent in developing countries.

Third World Forum (TWF)
P.O. Box 43
Orman, Cairo
Egypt

Phone: (20 2) 3488092
C: TRIFORUM
Fax: (20 2) 3480668
Founded in 1975. A network of social and other scientists and intellectuals in developing countries that works for "self-reliant, needs-oriented, endogenous development" through exchanging views, devising policy options, stimulating research, and seeking to influence policy-making.

Union of European Foresters (UEF)
Bijchen 6
D-7601 Ohlsbach
Germany
Phone: (49 7803) 40174
Founded in 1965. Objective is to promote all aspects of forests, forestry, and professional foresters.

Union of International Associations (UIA)
Rue Washington 40
B-1050 Brussels
Belgium
Phone: (32 2) 640 41 09
Tx: 65080 INAC B
Fax: (32 2) 649 32 69
Founded in 1907. Objective is to contribute to a universal order based on principles of human dignity, solidarity of peoples, and freedom of communication.

University of South Pacific—Institute of Natural Resources (INR)
P.O. Box 1168
Suva, Fiji
Phone: (679) 313900
Tx: 2712 fj
C: INR
Fax: (679) 300373
Founded in 1977. Objective is to conduct short training courses and undertake research and consulting in energy and the environment.

Wetland Link International (WLI)
c/o Wildfowl and Wetlands Trust
Slimbridge
Glos GL2 7BT
England
Phone: (44 453) 890333
Fax: (44 453) 890827
Founded in 1990. Promotes and coordinates the exchange of information and expertise among public wetland conservation education centers in various countries.

Wildlife Preservation Trust International
34th Street and Girard Avenue
Philadelphia, PA 19104
Phone: (215) 222-3636
Fax: (215) 222-2191
Founded in 1971. Promotes the preservation of endangered species through captive propagation, education, and field work.

Without Cruelty International (BWC)
57 King Henry's Walk
London NI 4NH
England
Phone: (44 71) 254 2929
Fax: (44 71) 923 4702
Founded in 1958. Promotes awareness of animal exploitation.

Women, Environment and Development Network

Women's Environment and Development Organization (WEDO)
845 Third Avenue, 15th Floor
New York, NY 10022
Phone: (212) 759-7982
Fax: (212) 759-8647
Promotes and advocates for women's issues worldwide.

World Association of Girl Guides and Girl Scouts (WAGGGS)
Olave Center, 12 c Lyndhurst Road
London NW3 5PQ
England
Phone: (44 71) 794 1181
C: WORLDBURO NW3
Fax: (44 71) 431 3764
Founded in 1928. Conducts and promotes environmental action projects and public awareness campaigns.

World Association of Soil and Water Conservation (WASWC)
317 Marvin Avenue
Volga, SD 57071
Phone: (605) 627-9309 or (218) 864-8506
Founded in 1982. Cooperates with the international Soil Conservation Organization and individuals interested in soil conservation worldwide.

World Blue Chain: For the Protection of Animals and Nature
(La Chaine Bleue Mondiale)
Avenue de Visé 39
B- 1 170 Brussels
Belgium
T: (32 2) 673 52 30
Founded in 1962. Promotes worldwide education to protect animals against bad treatment and cruelty and to protect nature and the environment.

World Commission on Environment and Development
(Brundtland Commission) (WCED)
Founded in 1983. WCED was charged with proposing long-term environmental strategies for achieving sustainable development; recommending ways in which concern for the environment may be translated into greater international cooperation; considering ways in which the international community can deal more effectively with environmental concerns; and helping to define shared perceptions of long-term environmental issues.

World Conservation Monitoring Centre (WCMC)
219c Huntingdon Road
Cambridge CB3 ODL
England
Phone: (44 223) 277314
Tx: 817036
C: REDBOOK
Fax: (44 223) 277136
WCMC, a joint activity of IUCN, WWF, and UNEP, gathers, manages and disseminates data on species, habitats, and sites through an extensive network of contacts in every country in the world. It also promotes the establishment of information networks to improve the flow and exchange of conservation data, including setting up data centers in developing countries.

The World Conservation Union
(aka International Union for Conservation of Nature and Natural Resources (IUCN)
Rue de Mauverney 28
CH-1196 Gland
Switzerland
Phone: (41 22) 999-0001
Tx: 419624 iucn ch
C: IUCNATURE
Fax: (41 22) 999-0002
Founded in 1948. IUCN, an independent international organization, is a union of sovereign states, governmental agencies, and nongovernmental organizations concerned with the initiation and promotion of scientifically based actions to ensure the perpetuation of man's natural environment.

World Council of Churches (WCC)
P.O. Box 2100
CH-1211 Geneva 2
Switzerland
Phone: (41 22) 791-611 1
Tx: 415 730 oik ch
Fax: (41 22) 791-0361
Founded in 1948. Promotes cooperation and facilitates common action by churches. The WCC Commission on the Churches' Participation in Development (CCPD) provides financial assistance to development projects.
World Council of Indigenous Peoples (WCIP)
555 King Edward Avenue

Ottawa, Ontario KIN 6N5
Canada
Phone: (I 613) 230-9030
Tx: 0533338
Fax: (I 613) 230-9340
Founded in 1975. Promotes protection of the rights and cultural interests of indigenous people.

World Federation of Engineering Organizations
(Federation Modiale des Organisations d'Ingénieurs) (WFEO)
1/7 Great George Street
London SWIP 3AA
England
T: (44 71) 222 7512
Tx: 935637 iceas g
Fax: (44 71) 222 0812
Founded in 1968. Objectives are to advance engineering as a profession in the interests of the world community; to foster cooperation between engineering organizations throughout the world; and to undertake special projects through cooperation between members in cooperation with other international bodies.

World Foundation for Environment and Development (WFED)
c/o Dr. Jon Martin Trolldalen
Department of Geography
University of Oslo
P.O. Box 1042
Blindern
N-0316 Oslo 3
Norway
Phone: (47 2) 85 59 33
Objective is to encourage and undertake collaborative initiatives to advance the peaceful resolution of international environmental conflicts.

World Future Studies Federation (WFSF)
c/o Social Science Research Institute
University of Hawaii
Honolulu, HI 96822
Phone: (808) 956-6601
Tx: 7238962
Fax: (808) 956-2889
Founded in 1973. Provides a forum for the exchange of information and opinions to stimulate cooperative research activities in all fields of futures research.

World Institute of Ecology and Cancer (WIEC)
Rue des Fripiers 24 Bis
B-1000 Brussels
Belgium
Phone: (32 2) 219 08 30
Conducts research activities concerning the environmental causes of cancer.

World Leisure and Recreation Association (WLRA)
P.O. Box 309, Sharbot Lake
Ontairo KOH 2PO,
Canada
Phone: (613) 2798-3173
Fax: (613) 2793372
Founded in 1956. Provides a worldwide forum for discussion of environmental issues related to leisure and recreation.

World Organization of the Scout Movement (WOSM)
P.O. Box 241
CH-1211 Geneva 4
Switzerland
Phone: (41 22) 320 42 33
C: Worldscout, Geneva
Tx: 428 139 wbs ch
Fax: (41 22) 781 20 53
Founded in 1920. Promotes and advises national scout organizations in more than 150 countries and territories.

World Pheasant Association (WPA)
P.O. Box 5, Lower Basildon
Reading RG8 9PF
England
Phone: (44 734) 845140
Fax: (44 734) 843369
Founded in 1975. Objective is to promote the conservation of all species of the order Galliformes.

World Resources Institute (WRI)

World Society for Ekistics (WSE)
24 Strat. Syndesmou Street
106 73 Athens, Greece
Phone: (30 1) 3623 216
Tx: 215227
Fax: (30 1) 3633 395
Founded in 1965. Promotes the development of knowledge and ideas about human settlements.

World Society for the Protection of Animals
Park Place, 10 Lawn Lane
London SW8 IUD
England
Phone: (44 71) 793-0540
Fax: (44 71) 7930208
Founded in 1981. Promotes the protection and conservation of animals worldwide.

World Sustainable Agriculture Association (WSAA)
8554 Melrose Avenue
West Hollywood, CA 90069
Phone: (310) 657-7202
Fax: (310) 657-3884
Founded in 1991. Promotes the well-being of humankind in harmony with nature through worldwide promotion of sustainable agriculture, holistic health, and the arts and humanities.

World Underwater Federation (Confederation Mondiale des Activités Subaquatiques) (CMAS)
47 Rue du Commerce
F-75015 Paris, France
Phone: (33 1) 45 75 42 75
Tx: 205734 f
Fax: (33 1) 45 77 11 04
Founded in 1959. CMAS is active in the protection and monitoring of aquatic environments.

WorldWIDE
P.O. Box 40885
Washington, DC 20016
Phone: (202) 347-1514
Fax: (202) 347-1524
Founded in 1981. Goals are to establish a worldwide network; educate the public and policymakers about the vital links between women, natural resources, and sustainable development; ensure that women and their environmental perceptions are incorporated in the design and implementation of development policies; and support and enhance the capacities of women.

World Wide Fund for Nature (WWF)
Avenue de Mont-Blanc
Ch-1196 Gland
Switzerland
Phone: (41 22) 64 91 1 1
Tx: 419624
Fax: (41 22) 64 53 58
Founded in 1961. WWF works through field work, policy development and lobbying, education and training, public awareness campaigns, and support for other organizations.

World Wildlife Fund

WorldWise
401 San Miguel Way
Sacramento, CA 95819
Phone: (916) 457-0433
Tx: 6503859849
Fax: (916) 739-6951
Founded in 1991. Provides grass roots campaign for reform of the "project-evaluation" processes of the multilateral development banks.

Xerces Society
10 S.W. Ash Street
Portland, OR 97204
Phone: (503) 222-2788
Fax: (503) 222-2763
Founded in 1971. Goal is to protect invertebrates as the major component of biological diversity.

Youth and Environment Europe (YEE)
Oude Gracht 42
NL-3511 AR Utrecht

Netherlands
Phone/Fax: (31 3) 311537

Youth for Development and Cooperation (YDC)
Overschiestraat 9
NL-1062 HN Amsterdam
Netherlands
Phone: (31 2) 142510
Fax: (31 2) 175545
Promotes protection of the environment, third world development, social justice, and international cooperation.

APPENDIX VII

GLOBAL AVAILABILITY OF WATER BY REGION

Region	Annual Internal Renewable Water Resources: Total (Thousands of Cubic Kilometers)	Annual Internal Renewable Water Resources: Per Capita (Thousands of Cubic Meters)	Percentage of Population Living in Countries with Scarce Annual per Capita Resources: Less Than 1,000 Cubic Meters	Percentage of Population Living in Countries with Scarce Annual per Capita Resources: 1,000–2,000 Cubic Meters
Sub-Saharan Africa and South Africa	3.8	7.1	8	16
East Asia and the Pacific	9.3	5.3	<1	6
South Asia	4.9	4.2	0	0
Eastern Europe and Former USSR	4.7	11.4	3	19
Other Europe	2.0	4.6	6	15
Middle East and North Africa	0.3	1.0	53	18
Latin America and the Caribbean	10.6	23.9	<1	4
Canada and the United States	5.4	19.4	0	0
World	40.9	7.7	4	8

Sources: World Resources Institute Data; World Bank Data 1994.

APPENDIX VIII

SELECTED WATER QUALITY INDICATORS FOR VARIOUS RIVERS

Country	River, City	*Dissolved Oxygen*				*Fecal Coliform*			
		ANNUAL MEAN CONCENTRATION (MILLIGRAMS PER LITER)			Average Annual Growth Rate for Series (percent)	ANNUAL MEAN CONCENTRATION (NUMBER PER 100-MILLILITER SAMPLE)			Average Annual Growth Rate for Series (percent)
		1979–82	1983–86	1987–90		1979–82	1983–86	1987–90	
Low-income									
Bangladesh	Karnaphuli	5.7	6.1	. .	−1.1 (5)	. .	. .	. .	. . (3)
Bangladesh	Meghna	6.5	7.0	. .	2.6 (5)	3,133	700	. .	−35.1 (5)
China	Pearl, Hong Kong	7.6	7.8	7.8	0.4 (11)	519	563	174	−14.4 (10)
China	Yangtze, Shanghai	8.3	8.3	8.2	−0.1 (11)	316	464	731	10.6 (11)
China	Yellow, Beijing	9.8	9.7	9.8	−0.1 (11)	711	1,337	1,539	9.8 (11)
India	Cauveri, d/s from KRS Reservoir	7.2	7.6	7.3	0.8 (9)	51	681	445	63.8 (9)
India	Cauveri, Satyagalam	7.0	7.3	7.5	1.1 (9)	10	684	920	121.8 (9)
India	Godavari, Dhalegaon	6.5	6.6	6.7	0.3 (9)	. .	. .	. .	. . (0)
India	Godavari, Mancherial	8.0	8.0	7.3	−1.1 (9)	5	5	8	19.7 (7)
India	Godavari, Polavaram	7.2	7.2	6.9	0.0 (8)	4	2	4	−3.8 (7)
India	Sabarmati, Dharoi	9.4	9.1	8.9	0.0 (9)	248	222	220	−15.4 (8)
India	Subarnarekha, Jamshedpur	8.0	7.9	7.5	−0.2 (9)	659	4,513	2,800	89.0 (9)
India	Subarnarekha, Ranchi	6.7	4.0	5.3	−6.2 (9)	1,239	7,988	3,100	70.5 (9)
India	Tapti, Burhanpur	7.5	6.9	6.1	−2.3 (9)	. .	110	130	−23.2 (4)
India	Tapti, Nepanagar	7.2	7.0	7.0	−0.6 (9)	. .	19	163	76.0 (4)
Pakistan	Chenab, Gujra Branch	6.2	6.8	7.1	1.8 (10)	436	463	446	−1.7 (10)
Pakistan	Indus, Kotri	7.6	7.2	2.6	−13.6 (11)	105	121	78	−3.4 (11)
Pakistan	Ravi, d/s from Lahore	6.8	5.7	6.3	−1.4 (12)	378	746	555	−2.4 (10)
Pakistan	Ravi, u/s from Lahore	7.2	6.7	7.0	−0.8 (12)	275	392	249	−6.6 (10)
Sudan	Blue Nile	7.3	8.2	. .	3.3 (7)	. .	. .	. .	. . (0)

Country	River, City	Dissolved Oxygen: Annual Mean Concentration (milligrams per liter) 1979–82	1983–86	1987–90	Dissolved Oxygen: Average Annual Growth Rate for Series (percent)	Fecal Coliform: Annual Mean Concentration (number per 100-milliliter sample) 1979–82	1983–86	1987–90	Fecal Coliform: Average Annual Growth Rate for Series (percent)
Middle-income									
Argentina	de la Plata, Buenos Aires	7.6	7.5	. .	0.0 (8)	828	230	. .	−23.1 (8)
Argentina	Paraná Corrientes	8.1	8.0	8.1	0.1 (10)	185	146	111	−6.6 (10)
Brazil	Guandu, Tomada d'Agua	8.1	7.8	7.7	−0.7 (11)	1,202	2,452	6	−47.0 (8)
Brazil	Paraiba, Aparecida	6.0	6.1	6.0	−0.4 (7)	13,950	9,800	6,075	−11.5 (7)
Brazil	Paraiba, Barra Mansa	7.4	7.6	7.8	0.4 (11)	8,003	8,100	8	−33.4 (7)
Chile	Maipo, el Manzano	12.9	13.2	10.8	−1.4 (10)	871	705	775	5.3 (8)
Chile	Mapocho, Los Almendros	11.8	12.1	10.0	−1.7 (10)	2	2	5	8.0 (8)
Colombia	Cauca Juanchito	. .	5.2	4.8	1.0 (5)	. .	10,000	10,000	0.0 (4)
Ecuador	San Pedro	7.7	7.8	. .	−0.1 (5)	80,000	30,603	. .	−31.5 (4)
Fiji	Waimanu	7.6	7.8	8.0	0.5 (9)	600	1,605	. .	8.1 (7)
Hungary	Danube	9.4	10.4	9.9	1.7 (10)	3,419	3,075	3,750	1.2 (10)
Korea	Han	. .	10.5	10.4	−0.2 (8)	. .	8	12	14.4 (8)
Malaysia	Kinta	6.8	7.5	8.3	2.9 (7)	. .	. .	. .	. . (0)
Malaysia	Klang	3.0	3.3	2.8	−1.1 (9)	. .	. .	. .	. . (1)
Malaysia	Linggi	3.4	3.6	3.7	0.9 (10)	. .	. .	. .	. . (0)
Malaysia	Muda	7.3	7.2	6.3	−1.3 (8)	. .	. .	. .	. . (0)

Country	River, City	Dissolved Oxygen: Annual Mean Concentration (milligrams per liter) 1979–82	1983–86	1987–90	Dissolved Oxygen: Average Annual Growth Rate for Series (percent)	Fecal Coliform: Annual Mean Concentration (number per 100-milliliter sample) 1979–82	1983–86	1987–90	Fecal Coliform: Average Annual Growth Rate for Series (percent)
Middle-income (*Continued*)									
Mexico	Atoyac	3.5	1.7	0.3	−47.5 (9)	157,500	105,000	916,667	23.9 (7)
Mexico	Balsas	7.6	6.3	6.8	−1.9 (10)	1,558	26,333	130,000	95.4 (8)
Mexico	Blanco	5.0	3.4	4.1	−3.7 (9)	21,717	39,500	12,150	1.8 (8)
Mexico	Colorado	7.9	8.7	8.2	1.4 (9)	277	58	37	−28.7 (7)
Mexico	Lerma	0.3	0.4	0.5	−18.6 (10)	192,250	165,000	67	5.7 (7)
Mexico	Panuco	7.7	8.1	8.3	0.7 (11)	110	201	. .	−27.8 (6)
Panama	Aguas Claras	7.9	8.2	. .	0.4 (7)	219	143	. .	−14.4 (6)
Panama	San Felix	8.2	8.0	. .	−1.0 (7)	850	753	. .	−6.2 (6)
Philippines	Cagayan	7.8	7.9	8.1	0.3 (11)	. .	. .	. .	. . (3)
Portugal	Tejo, Santarem	8.9	8.6	8.4	−0.7 (9)	2,252	4,163	4,225	24.6 (9)
Thailand	Chao Phrya, d/s from Nakhon Sawan	6.3	6.3	. .	0.2 (8)	1,093	1,745	. .	47.7 (7)
Thailand	Prasak, Kaeng Khoi	6.6	7.7	. .	8.0 (5)	596	2,724	. .	9.9 (8)
Turkey	Porsuk, Agackoy	9.0	9.1	9.2	0.7 (9)	. .	. .	. .	. . (1)
Turkey	Sekarya, Adetepe	9.2	8.7	8.9	−0.3 (8)	. .	. .	. .	. . (1)
Uruguay	de la Plata, Colonia	. .	. .	. .	7.1 (3)	. .	453	93	54.6 (4)
Uruguay	Uruguay Bella Unión	. .	7.9	8.4	−1.4 (4)	. .	200	1,100	66.9 (4)

Country	River, City	Dissolved Oxygen: Annual Mean Concentration (milligrams per liter) 1979–82	1983–86	1987–90	Dissolved Oxygen: Average Annual Growth Rate for Series (percent)	Fecal Coliform: Annual Mean Concentration (number per 100-milliliter sample) 1979–82	1983–86	1987–90	Fecal Coliform: Average Annual Growth Rate for Series (percent)
High-income									
Australia	Murray	10.0	9.4	9.1	1.0 (6)	. .	. .	. .	. . (0)
Australia	Murray, Mannum	7.1	8.2	8.6	2.4 (8)	33	103	80	15.8 (8)
Belgium	Escaut, Bleharies	5.7	6.2	5.9	1.1 (11)	76	579	867	40.8 (11)
Belgium	Meuse, Heer/Agimont	10.5	10.8	11.3	0.8 (11)	30	1,391	1,700	69.7 (11)
Belgium	Meuse, Lanaye Ternaaien	9.2	8.4	8.9	−0.7 (11)	147	5,233	7,100	78.2 (11)
Japan	Kiso, Asahi	10.0	10.6	11.7	1.7 (11)	300	400	216	−4.1 (11)
Japan	Kiso, Inuyama	10.8	10.5	10.8	−0.2 (10)	610	491	600	−2.0 (10)
Japan	Kiso, Shimo-Ochiai	11.2	11.1	11.4	0.3 (10)	546	443	353	−6.0 (10)
Japan	Shinano, Zuiun Bridge	10.1	10.3	10.3	0.2 (10)	290	346	193	−3.0 (10)
Japan	Tone, Tone-Ozeki	10.0	9.9	10.4	0.5 (10)	521	593	618	3.7 (10)
Japan	Yodo, Hirakata Bridge	8.7	8.4	8.4	−0.4 (11)	72,000	70,333	. .	93 (7)
Netherlands	Ijssel, (arm of Rhine)	8.7	7.9	. .	−3.3 (6)	9,833	2,050	. .	−43.0 (5)
Netherlands	Rhine, German frontier	8.5	8.0	. .	−2.6 (6)	17,633	10,500	. .	−11.8 (5)
United Kingdom	Thames	9.9	10.3	9.1	0.2 (8)	. .	. .	. .	. . (0)
United States	Delaware, Trenton, N.J.	11.1	10.6	. .	−2.5 (7)	74	197	. .	−4.0 (7)
United States	Hudson, Green Island, N.Y.	9.8	12.1	. .	4.2 (7)	941	792	. .	−7.4 (7)
United States	Mississippi, Vicksburg, Miss.	8.4	8.3	. .	−0.2 (7)	435	1,473	. .	40.2 (7)

Source: World Bank.

APPENDIX IX

GLOBAL ANNUAL AVERAGE CARBON DIOXIDE (CO_2) CONCENTRATION VALUES—1997 DATA (in parts per million by volume)

Station	Annual Average (ppmv)
Barrow, Alaska, U.S.A. On the coast of the Arctic Ocean 71°19′ N, 156°36′ W, 11m above MSL	365.07ppm
Mauna Loa, Hawaii, U.S.A. Barren lava field of an active volcano 19°32′ N, 155°35′ W, 3397m above MSL	363.82ppm
Samoa (Cape Matatula), U.S. Territory Pacific Ocean rocky coastal promontory 14°15′ S, 170°34′ W, 30m above MSL	362.43ppm
South Pole, Antarctica Ice- and snow-covered plateau 89°59′ S, 24°48′ W, 2810m above MSL	361.25ppm

Source: Data contributed by C.D. Keeling and T.P. Whorf (Scripps Institution of Oceanography) The Carbon Dioxide Information Analysis Center (CDIAC) Oak Ridge National Laboratory, Oak Ridge, TN.

APPENDIX X

CHANGES IN LAND USE FOR LOW- AND MIDDLE-INCOME COUNTRIES

Country Group	Land Area, 1989 (thousands of square kilometers)	*Share of Total Land Area, 1989 (percent)*				*Average Annual Growth Rate, 1965–89 (percent)*			
		Agricultural	Permanent Pasture	Forest and Woodland	Other	Agricultural	Permanent Pasture	Forest and Woodland	Other
Low-income	36,396	13	27	25	36	0.2	0.0	−0.4	0
China and India	12,264	22	27	16	36	0.0	0.0	−0.4	0
Other low-income	24,132	9	27	29	35	0.5	0.0	−0.4	0
Mozambique	784	4	56	18	22	0.5	0.0	−0.8	0
Tanzania	886	6	40	46	8	0.9	0.0	−0.3	1
Ethiopia	1,101	13	41	25	22	0.4	−0.1	−0.4	0
Somalia	627	2	69	14	15	0.5	0.0	−0.1	0
Nepal	137	19	15	18	48	1.6	0.9	0.0	−0
Chad	1,259	3	36	10	52	0.5	0.0	−0.6	0
Bhutan	47	3	6	55	36	1.3	0.3	0.4	−0
Lao PDR	231	4	3	55	37	0.4	0.0	−0.7	1
Malawi	94	26	20	40	15	0.8	0.0	−1.1	4
Bangladesh	130	71	5	15	9	0.1	0.0	−0.4	0
Burundi	26	52	36	3	10	1.1	1.9	1.3	−5
Zaire	2,268	3	7	77	13	0.5	0.0	−0.2	1
Uganda	200	34	9	28	29	1.6	0.0	−0.5	−0
Madagascar	582	5	58	27	9	1.6	0.0	−0.9	2
Sierra Leone	72	25	31	29	15	1.1	0.0	−0.2	−1
Mali	1,220	2	25	6	68	1.1	0.0	−0.4	0
Nigeria	911	34	44	13	8	0.3	0.0	−1.9	4
Niger	1,267	3	7	2	88	2.3	−0.7	−2.2	0
Rwanda	25	47	19	23	12	2.7	−2.8	−0.5	−0
Burkina Faso	274	13	37	24	26	2.1	0.0	−0.8	0
India	2,973	57	4	22	17	0.2	−0.8	0.3	−0
Benin	111	17	4	32	47	1.0	0.0	−1.4	0

Country Group	Land Area, 1989 (thousands of square kilometers)	*Share of Total Land Area, 1989 (percent)*				*Average Annual Growth Rate, 1965–89 (percent)*			
		Agricultural	Permanent Pasture	Forest and Woodland	Other	Agricultural	Permanent Pasture	Forest and Woodland	Other
China	9,291	10	34	13	42	−0.3	0.0	−0.8	0
Haiti	28	33	18	1	48	0.8	−1.2	−2.3	0
Kenya	570	4	67	4	25	1.0	0.0	−0.8	0
Pakistan	771	27	6	5	62	0.4	0.0	1.9	−0
Ghana	230	12	22	35	31	−0.3	0.0	−0.8	1
Central African Republic	623	3	5	57	34	0.5	0.0	0.0	0
Togo	54	27	33	30	11	0.6	0.0	−0.6	0
Zambia	743	7	40	39	14	0.3	0.0	−0.2	0
Guinea	246	3	25	60	12	0.4	0.0	−0.4	2
Sri Lanka	65	29	7	27	37	0.0	0.8	−0.3	0
Mauritania	1,025	0	38	5	57	−1.6	0.0	0.3	0
Lesotho	30	11	66	. .	24	−1.0	−0.4	. .	1
Indonesia	1,812	12	7	63	19	0.9	−0.3	−0.5	1
Honduras	112	16	23	30	31	0.9	0.9	−1.9	1
Egypt, Arab Rep.	995	3	. .	0	97	−0.6	. .	0.0	0
Afghanistan	652	12	46	3	39	0.1	0.0	−0.2	0
Cambodia	177	17	3	76	4	0.1	0.0	0.0	−0
Liberia	96	4	59	18	19	0.1	0.0	−1.2	1
Myanmar	658	15	1	49	35	−0.2	−0.1	0.0	0
Sudan	2,376	5	41	19	34	0.5	0.0	−0.6	0
Viet Nam	325	20	1	30	49	0.5	0.8	−2.3	1

Country Group	Land Area, 1989 (thousands of square kilometers)	*Share of Total Land Area, 1989 (percent)*				*Average Annual Growth Rate, 1965–89 (percent)*			
		Agricultural	Permanent Pasture	Forest and Woodland	Other	Agricultural	Permanent Pasture	Forest and Woodland	Other
Middle-income	40,684	10	29	33	29	0.7	0.1	−0.4	0
Lower-middle-income	22,141	10	31	28	31	0.6	−0.1	−0.5	0
Bolivia	1,084	3	25	51	21	3.1	−0.2	−0.3	0
Zimbabwe	387	7	13	50	30	1.1	0.0	−0.4	0
Senegal	193	27	30	31	12	0.7	0.0	−0.6	0
Côte d'Ivoire	318	12	41	24	24	1.4	0.0	−2.4	4
Philippines	298	27	4	35	34	0.7	2.0	−2.1	3
Dominican Republic	48	30	43	13	14	1.5	0.0	−0.3	−2
Papua New Guinea	453	1	0	84	15	0.8	0.5	−0.1	0
Guatemala	108	17	13	35	35	1.0	0.8	−1.3	1
Morocco	446	21	47	18	15	0.9	0.9	0.2	−2
Cameroon	465	15	18	53	14	1.0	−0.3	−0.4	1
Ecuador	277	10	18	40	32	0.1	4.5	−1.8	1
Syrian Arab Rep.	184	30	43	4	23	−0.5	0.4	0.7	−0
Congo	342	0	29	62	8	1.0	0.0	−0.1	0
El Salvador	21	35	29	5	30	0.8	0.0	−2.9	−0
Paraguay	397	6	52	36	6	5.0	1.6	−1.4	−2
Peru	1,280	3	21	54	22	1.7	0.0	−0.4	0
Jordan	89	4	9	1	86	0.9	0.0	1.2	−0
Colombia	1,039	5	39	49	7	0.3	0.7	−0.6	0
Thailand	511	43	2	28	27	2.4	3.5	−2.6	0
Tunisia	155	30	19	4	47	0.3	0.9	1.5	−0
Jamaica	11	25	18	17	40	0.2	−1.1	−0.5	0
Turkey	770	36	11	26	26	0.2	−1.2	0.0	0
Romania	230	45	19	28	8	0.0	0.0	0.0	−0
Poland	304	48	13	29	10	−0.2	−0.3	0.3	0

Source: World Development Report 1992—Development and Environment, World Bank Report.

APPENDIX XI

NATIONALLY PROTECTED AREAS

Country Group	*All Nationally Protected Areas (thousand sqaure kilometers)*		*Number of Protected Areas*		*Protected Areas as a Share of Total Land Area (percent)*		*Share of Protected Areas Totally Protected (percent)*		*Share of Protected Areas Partially Protected (percent)*	
	1972	1990	1972	1990	1972	1990	1972	1990	1972	1990
Low-income	592	1,441	361	1,407	1.6	3.8	59	46	41	54
China and India	27	411	84	736	0.2	3.2	12	11	88	89
Other low-income	565	1,031	277	671	2.3	4.1	61	61	39	39
Middle-income	778	2,215	691	1,839	1.9	5.3	70	53	30	47
Lower-middle-income	623	1,316	377	975	2.7	5.8	71	65	29	35
Upper-middle-income	156	899	314	864	0.8	4.8	67	35	33	65
Low- and middle-income	1,370	3,656	1,052	3,246	1.7	4.6	65	50	35	50
Sub-Saharan Africa	790	1,105	251	379	3.4	4.8	65	65	35	35
East Asia and the Pacific	58	611	150	857	0.4	3.9	38	37	62	63
South Asia	32	198	110	469	0.6	3.8	17	34	83	66
Europe	16	77	144	411	0.7	3.6	48	20	52	80
Middle East and North Africa	128	427	50	126	1.1	3.7	80	38	20	62
Latin American and the Caribbean	293	1,173	238	797	1.4	5.8	72	53	28	47
Other economies	75	247	109	231	0.3	1.1	95	97	5	3
High-income	988	3,412	1,840	3,632	2.9	10.2	49	67	51	33
OECD members	986	2,423	1,820	3,581	3.2	7.8	48	54	52	46
Other	2	989	20	51	0.1	41.5	71	100	29	0
World[a]	2,434	7,354	3,012	7,152	1.6	4.9	59	60	41	40

[a]Includes countries not elsewhere specified and some economies with populations under 30,000.

APPENDIX XII

CENTRAL GOVERNMENT EXPENDITURES

	Defense		*Education*		*Health*		*Housing, etc., Soc. Sec., Welfare*		*Economic Services*		*Other*		*Total Expenditure (% of GNP)*		*Overall Surplus/Deficit (% of GNP)*	
	1980	1992	1980	1992	1980	1992	1980	1992	1980	1992	1980	1992	1980	1992	1980	1992
Low-income economies Excluding China and India																
1 Mozambique	. .	. .	. .	. .	. .	. .	. .	. .	. .	. .	. .	. .	. .	. .	. .	. .
2 Ethiopia	. .	. .	10.1	. .	3.7	. .	5.4	. .	23.8	. .	. .	. .	23.4	. .	−4.5	. .
3 Tanzania	9.2	. .	13.3	. .	6.0	. .	2.5	. .	42.9	. .	26.1	. .	28.8	. .	−8.4	. .
4 Sierra Leone	*4.1*	*9.9*	*14.9*	*13.3*	*9.1*	*9.6*	*3.6*	*3.1*	. .	*29.0*	*68.3*	*35.2*	29.8	19.6	−13.2	−6.2
5 Nepal	6.7	*5.9*	9.9	*10.9*	3.9	*4.7*	1.7	*6.8*	58.8	*43.0*	19.1	*28.8*	14.2	*18.7*	−3.0	*−6.3*
6 Uganda	25.2	. .	14.9	. .	5.1	. .	4.2	. .	11.1	. .	39.5	. .	6.1	. .	−3.1	. .
7 Bhutan	*0.0*	*0.0*	*12.8*	*10.7*	*5.0*	*4.8*	*4.9*	*8.2*	*56.8*	*48.2*	*20.5*	*28.2*	40.6	*40.9*	0.9	*−2.5*
8 Burundi	. .	. .	. .	. .	. .	. .	. .	. .	. .	. .	. .	. .	21.7	. .	−3.9	. .
9 Malawi	12.8	*4.8*	9.0	*10.4*	5.5	*7.8*	1.6	*4.2*	43.7	*35.6*	27.3	*37.2*	37.6	*26.6*	−17.3	*−1.7*
10 Bangladesh	9.4	. .	11.5	. .	6.4	. .	5.3	. .	46.9	. .	20.4	. .	10.0	. .	2.5	. .
11 Chad	. .	. .	. .	. .	. .	. .	. .	. .	. .	. .	. .	. .	. .	*32.0*	. .	*−7.5*
12 Guinea-Bissau	. .	. .	. .	. .	. .	. .	. .	. .	. .	. .	. .	. .	. .	. .	. .	. .
13 Madagascar	. .	*7.5*	. .	*17.2*	. .	*6.6*	. .	*1.5*	. .	*35.9*	. .	*31.2*	. .	*16.1*	. .	*−5.9*
14 Lao PDR	. .	. .	. .	. .	. .	. .	. .	. .	. .	. .	. .	. .	. .	. .	. .	. .
15 Rwanda	13.1	. .	18.8	. .	4.5	. .	4.1	. .	41.4	. .	18.0	. .	14.3	26.1	−1.7	−7.2
16 Niger	3.8	. .	18.0	. .	4.1	. .	3.8	. .	32.4	. .	38.0	. .	18.7	. .	−4.8	. .
17 Burkina Faso	17.0	. .	15.5	. .	5.8	. .	7.6	. .	19.3	. .	34.8	. .	14.1	. .	0.3	. .
18 India	19.8	15.0	1.9	2.1	1.6	1.6	4.3	5.7	24.2	18.6	48.3	57.0	13.2	16.8	−6.5	−4.9
19 Kenya	16.4	*9.2*	19.6	*20.1*	7.8	*5.4*	5.1	*3.4*	22.7	*18.1*	28.2	*43.8*	26.1	*30.7*	−4.6	*−2.8*
20 Mali	11.0	. .	15.7	. .	3.1	. .	3.0	. .	11.2	. .	46.0	. .	21.6	. .	−4.7	. .
21 Nigeria	. .	. .	. .	. .	. .	. .	. .	. .	. .	. .	. .	. .	. .	. .	. .	. .
22 Nicaragua	11.0	. .	11.6	. .	14.6	. .	7.4	. .	20.6	. .	34.9	. .	32.3	*39.3*	−7.3	*−17.7*
23 Togo	*7.2*	. .	*16.7*	. .	*5.3*	. .	*12.0*	. .	*25.2*	. .	*33.7*	. .	31.9	. .	−2.0	. .
24 Benin	. .	. .	. .	. .	. .	. .	. .	. .	. .	. .	. .	. .	. .	. .	. .	. .
25 Central African Republi	*9.7*	. .	*17.6*	. .	*5.1*	. .	*6.3*	. .	*19.6*	. .	*41.7*	. .	21.9	. .	−3.5	. .

	Defense		*Education*		*Health*		*Housing, etc., Soc. Sec., Welfare*		*Economic Services*		*Other*		*Total Expenditure (% of GNP)*		*Overall Surplus/Deficit (% of GNP)*	
	1980	1992	1980	1992	1980	1992	1980	1992	1980	1992	1980	1992	1980	1992	1980	1992
Low-income economies (*continued*) Excluding China and India																
26 Pakistan	30.6	*27.9*	2.7	*1.6*	1.5	*1.0*	4.1	*3.4*	37.2	*11.6*	23.9	*54.6*	17.7	*21.7*	−5.8	*−6.2*
27 Ghana	3.7	. .	22.0	. .	7.0	. .	6.8	. .	20.7	. .	39.8	. .	10.9	. .	−4.2	. .
28 China	. .	. .	. .	. .	. .	. .	. .	. .	. .	. .	. .	. .	. .	. .	. .	. .
29 Tajikistan	. .	. .	. .	. .	. .	. .	. .	. .	. .	. .	. .	. .	. .	. .	. .	. .
30 Guinea	. .	. .	. .	. .	. .	. .	. .	. .	. .	. .	. .	. .	. .	*23.1*	. .	*−3.9*
31 Mauritania	. .	. .	. .	. .	. .	. .	. .	. .	. .	. .	. .	. .	. .	. .	. .	. .
32 Sri Lanka	1.7	8.5	6.7	10.1	4.9	4.8	12.7	16.1	15.9	24.0	58.2	36.5	41.6	28.2	−18.4	−7.2
33 Zimbabwe	25.0	. .	15.5	. .	5.4	. .	7.8	. .	18.1	. .	28.2	. .	35.3	*34.8*	−11.1	*−6.7*
34 Honduras	. .	. .	. .	. .	. .	. .	. .	. .	. .	. .	. .	. .	. .	. .	. .	. .
35 Lesotho	*0.0*	*6.5*	*15.3*	*21.9*	*6.2*	*11.5*	*1.3*	*5.5*	*35.9*	*31.6*	*41.2*	*23.1*	22.7	*33.2*	−3.7	*−0.3*
36 Egypt, Arab Republic	11.4	. .	8.1	. .	2.4	. .	13.1	. .	7.2	. .	57.7	. .	53.7	. .	−12.5	. .
37 Indonesia	13.5	*6.8*	8.3	*9.8*	2.5	*2.8*	1.8	*2.0*	40.2	*29.6*	33.7	*49.1*	23.1	*19.2*	−2.3	*0.5*
38 Myanmar	21.9	*22.0*	10.6	*17.4*	5.3	*6.8*	10.6	*12.1*	33.7	*19.5*	17.9	*22.1*	15.9	*15.5*	1.2	*−5.0*
39 Somalia	. .	. .	. .	. .	. .	. .	. .	. .	. .	. .	. .	. .	. .	. .	. .	. .
40 Sudan	13.2	. .	9.8	. .	1.4	. .	0.9	. .	19.8	. .	54.9	. .	19.8	. .	−3.3	. .
41 Yemen, Republic of	. .	. .	. .	. .	. .	. .	. .	. .	. .	. .	. .	. .	. .	. .	. .	. .
42 Zambia	0.0	. .	11.4	. .	6.1	. .	3.4	. .	32.6	. .	46.6	. .	40.0	. .	−20.0	. .

	Defense		Education		Health		Housing, etc., Soc. Sec., Welfare		Economic Services		Other		Total Expenditure (% of GNP)		Overall Surplus/Deficit (% of GNP)	
	1980	1992	1980	1992	1980	1992	1980	1992	1980	1992	1980	1992	1980	1992	1980	1992
Middle-income economies																
Lower-middle-income																
43 Côte d'Ivoire	3.9	. .	16.3	. .	3.9	. .	4.3	. .	13.4	. .	58.1	. .	33.3	*31.2*	−11.4	*−3.7*
44 Bolivia	. .	9.8	. .	16.6	. .	8.2	. .	12.7	. .	16.1	. .	36.6	29.0	22.5	0.0	−2.3
45 Azerbaijan	. .	. .	. .	. .	. .	. .	. .	. .	. .	. .	. .	. .	. .	. .	. .	. .
46 Philippines	15.7	9.9	13.0	15.0	4.5	4.1	6.6	4.4	56.9	26.8	3.4	39.8	13.4	19.4	−1.4	−1.2
47 Armenia	. .	. .	. .	. .	. .	. .	. .	. .	. .	. .	. .	. .	. .	. .	. .	. .
48 Senegal	16.8	. .	23.0	. .	4.7	. .	9.5	. .	14.4	. .	31.6	. .	23.9	. .	0.9	. .
49 Cameroon	9.1	. .	12.4	. .	5.1	. .	8.0	. .	24.0	. .	41.4	. .	15.5	20.3	0.5	−2.2
50 Kyrgyz Republic	. .	. .	. .	. .	. .	. .	. .	. .	. .	. .	. .	. .	. .	. .	. .	. .
51 Georgia	. .	. .	. .	. .	. .	. .	. .	. .	. .	. .	. .	. .	. .	. .	. .	. .
52 Uzbekistan	. .	. .	. .	. .	. .	. .	. .	. .	. .	. .	. .	. .	. .	. .	. .	. .
53 Papua New Guinea	4.4	4.2	16.5	15.0	8.6	7.9	2.6	1.4	22.7	21.6	45.1	49.9	35.2	36.0	−2.0	−5.9
54 Peru	21.0	. .	15.6	. .	5.6	. .	0.0	. .	22.1	. .	35.7	. .	20.4	12.5	−2.5	−1.7
55 Guatemala	. .	. .	. .	. .	. .	. .	. .	. .	. .	. .	. .	. .	14.4	. .	−3.9	. .
56 Congo	*9.7*	. .	*11.0*	. .	*5.1*	. .	*7.0*	. .	*34.2*	. .	*33.0*	. .	54.6	. .	−5.8	. .
57 Morocco	17.9	*12.8*	17.3	*18.2*	3.4	*3.0*	6.5	*5.8*	27.8	*15.2*	27.1	*44.9*	34.2	*29.8*	−10.0	*−2.3*
58 Dominican Republic	7.8	*4.8*	12.6	*10.2*	9.3	*14.0*	13.8	*20.2*	37.1	*36.5*	19.3	*14.2*	17.5	*12.3*	−2.7	*0.6*
59 Ecuador	12.5	*12.9*	34.7	*18.2*	7.8	*11.0*	1.3	*2.5*	21.1	*11.8*	22.6	*43.6*	15.0	*15.9*	−1.5	*2.0*
60 Jordan	*25.3*	*26.7*	*7.6*	*12.9*	*3.7*	*5.2*	*14.5*	*15.1*	*28.3*	*10.7*	*20.6*	*29.5*	. .	*41.7*	. .	*−3.1*
61 Romania	. .	*10.3*	. .	*10.0*	. .	*9.2*	. .	*26.6*	. .	*33.0*	. .	*10.9*	. .	*37.0*	. .	*2.0*
62 El Salvador	8.8	16.0	19.8	12.8	9.0	7.3	5.5	4.7	21.0	19.4	36.0	39.7	17.6	11.2	−5.9	−0.8
63 Turkmenistan	. .	. .	. .	. .	. .	. .	. .	. .	. .	. .	. .	. .	. .	. .	. .	. .
64 Moldova	. .	. .	. .	. .	. .	. .	. .	. .	. .	. .	. .	. .	. .	. .	. .	. .
65 Lithuania	. .	. .	. .	. .	. .	. .	. .	. .	. .	. .	. .	. .	. .	. .	. .	. .
66 Bulgaria	. .	*5.6*	. .	*6.2*	. .	*4.8*	. .	*23.9*	. .	*46.6*	. .	*12.8*	. .	42.4	. .	−5.1
67 Colombia	6.7	. .	19.1	. .	3.9	. .	21.2	. .	27.1	. .	22.0	. .	13.5	. .	−1.8	. .
68 Jamaica	. .	. .	. .	. .	. .	. .	. .	. .	. .	. .	. .	. .	45.7	. .	−17.1	. .
69 Paraguay	12.4	*13.3*	12.9	*12.7*	3.6	*4.3*	19.2	*14.8*	18.9	*12.8*	33.0	*42.1*	9.8	*9.4*	0.3	*3.0*
70 Namibia	. .	*6.5*	. .	*22.2*	. .	*9.7*	. .	*14.8*	. .	*17.3*	. .	*29.5*	. .	*44.2*	. .	*−6.9*
71 Kazakhstan	. .	. .	. .	. .	. .	. .	. .	. .	. .	. .	. .	. .	. .	. .	. .	. .
72 Tunisia	12.2	5.4	17.0	17.5	7.2	6.6	13.4	18.6	27.8	22.5	22.4	29.3	32.5	32.8	−2.9	−2.6

Source: World Bank Report 1993.

APPENDIX XIII

DEVELOPMENT ASSISTANCE RECEIPTS

Net Disbursement of ODA from All Sources

	MILLIONS OF DOLLARS							Per Capita ($)	As Percentage
	1985	1986	1987	1988	1989	1990	1991	1991	of GNP 1991
Low-income economies	17,065	19,038	20,988	24,004	24,530	30,441	31,711	10.2	2.7
Excluding China and India	14,533	15,785	17,688	19,918	20,482	26,836	27,010	25.1	7.0
1 Mozambique	300	422	651	893	772	935	920	57.1	69.2
2 Ethiopia	710	636	634	970	752	1,014	1,091	20.6	16.5
3 Tanzania	484	681	882	982	920	1,141	1,076	42.7	33.8
4 Sierra Leone	65	87	68	102	100	65	105	24.7	13.9
5 Nepal	234	301	347	399	493	430	453	23.4	13.6
6 Uganda	180	198	280	363	443	551	525	31.1	20.5
7 Bhutan	24	40	42	42	42	48	64	43.8	25.4
8 Burundi	139	187	202	189	199	265	253	44.7	21.6
9 Malawi	113	198	280	366	412	481	495	56.2	22.6
10 Bangladesh	1,131	1,455	1,635	1,592	1,800	2,048	1,636	14.6	7.0
11 Chad	181	165	198	264	241	303	262	44.9	20.2
12 Guinea-Bissau	58	71	111	99	101	117	101	101.3	43.4
13 Madagascar	185	316	321	304	321	386	437	36.4	16.4
14 Lao PDR	37	48	58	77	140	152	131	30.8	12.7
15 Rwanda	180	211	245	252	232	293	351	49.1	21.5
16 Niger	303	307	353	371	296	391	376	47.6	16.2
17 Burkina Faso	195	284	281	298	272	336	409	44.1	14.8
18 India	1,592	2,120	1,839	2,097	1,895	1,524	2,747	3.2	1.1
19 Kenya	430	455	572	808	967	1,053	873	35.0	10.9
20 Mali	376	372	366	427	454	467	455	52.2	18.5

Net Disbursement of ODA from All Sources

	MILLIONS OF DOLLARS							Per Capita ($)	As Percentage
	1985	1986	1987	1988	1989	1990	1991	1991	of GNP 1991
21 Nigeria	1,032	59	69	120	346	250	262	2.6	0.8
22 Nicaragua	102	150	141	213	225	320	826	219.0	47.6
23 Togo	111	174	126	199	183	241	204	54.0	12.4
24 Benin	94	138	138	162	263	271	256	52.4	13.5
25 Central African Republic	104	139	176	196	192	244	174	56.4	13.6
26 Pakistan	769	970	879	1,408	1,129	1,149	1,226	10.6	2.7
27 Ghana	196	371	373	474	550	498	724	47.2	10.3
28 China	940	1,134	1,462	1,989	2,153	2,081	1,954	1.7	0.4
29 Tajikistan	. .	. .	. .	. .	. .	. .	. .	. .	. .
30 Guinea	115	175	213	262	346	296	371	62.6	11.7
31 Mauritania	207	225	185	184	242	202	208	102.9	18.4
32 Sri Lanka	468	570	502	598	547	674	814	47.2	9.0
33 Zimbabwe	237	225	294	273	265	340	393	39.2	6.0
34 Honduras	270	283	258	321	242	450	275	52.2	9.1
35 Lesotho	93	88	107	108	127	139	123	67.9	20.5
36 Egypt, Arab Republic	1,760	1,716	1,773	1,537	1,568	5,444	4,988	93.1	15.2
37 Indonesia	603	711	1,246	1,632	1,839	1,724	1,854	10.2	1.6
38 Myanmar	. .	. .	. .	. .	. .	. .	. .	. .	. .
39 Somalia	353	511	580	433	427	485	186	23.1	. .
40 Sudan	1,128	945	898	937	772	825	887	34.4	. .
41 Yemen, Rep.	392	328	422	304	370	405	313	25.0	. .
42 Zambia	322	464	430	478	392	486	884	110.2	. .

	Net Disbursement of ODA from All Sources								
	MILLIONS OF DOLLARS							Per Capita ($)	As Percentage
	1985	1986	1987	1988	1989	1990	1991	1991	of GNP 1991
Middle-income economies	9,057	9,470	10,487	9,680	10.062	15,457	15,535	16.4	0.7
Low-middle-income	6,817	7,875	8,680	8,179	8,408	13,152	13,453	24.4	1.8
43 Côte d'Ivoire	117	186	254	439	403	693	633	50.9	6.7
44 Bolivia	197	322	318	394	440	506	473	64.4	9.4
45 Azerbaijan	. .	. .	. .	. .	. .	. .	. .	. .	. .
46 Philippines	460	956	770	854	844	1,279	1,051	16.7	2.3
47 Amenia	. .	. .	. .	. .	. .	. .	. .	. .	. .
48 Senegal	289	567	641	569	650	788	577	75.7	10.2
49 Cameroon	153	224	213	284	458	431	501	42.2	4.3
50 Kyrgyz Republic	. .	. .	. .	. .	. .	. .	. .	. .	. .
51 Georgia	. .	. .	. .	. .	. .	. .	. .	. .	. .
52 Uzbekistan	. .	. .	. .	. .	. .	. .	. .	. .	. .
53 Papua New Guinea	257	263	322	380	339	416	397	100.1	10.5
54 Peru	316	272	292	272	305	395	590	26.9	2.7
55 Guatemala	83	135	241	235	261	203	197	20.8	2.1
56 Congo	69	110	152	89	91	214	133	56.7	4.9
57 Morocco	766	403	447	480	450	1,026	1,075	41.9	3.9
58 Dominican Republic	207	93	130	118	142	100	66	9.1	0.9
59 Ecuador	136	147	203	137	160	155	220	20.4	1.9
60 Jordan	538	564	577	417	273	884	905	247.1	22.2
61 Romania	. .	. .	. .	. .	. .	. .	. .	. .	. .
62 El Salvador	345	341	426	420	443	349	290	54.9	4.9

Net Disbursement of ODA from All Sources

	MILLIONS OF DOLLARS							Per Capita ($)	As Percentage
	1985	1986	1987	1988	1989	1990	1991	1991	of GNP 1991
63 Turkmenistan	. .	. .	. .	. .	. .	. .	. .	. .	. .
64 Moldova	. .	. .	. .	. .	. .	. .	. .	. .	. .
65 Lithuania	. .	. .	. .	. .	. .	. .	. .	. .	. .
66 Bulgaria	. .	. .	. .	. .	. .	. .	. .	. .	. .
67 Colombia	62	63	78	61	67	88	123	3.8	0.3
68 Jamaica	169	178	168	193	262	273	166	69.7	4.7
69 Paraguay	50	66	81	76	92	56	144	32.6	2.3
70 Namibia	6	15	17	22	59	123	184	124.1	8.2
71 Kazakhstan	. .	. .	. .	. .	. .	. .	. .	. .	. .
72 Tunisia	163	222	274	316	283	393	322	39.1	2.4

APPENDIX XIV

GLOBAL WATER AVAILABILITY

Country Group	Total Annual Internal Renewable Water Resources (cubic kilometers)	Total Annual Water Withdrawal (cubic kilometers)	Annual Withdrawal as a Share of Total Water Resources (percent)	Per Capita Annual Internal Renewable Water Resources, 1990 (cubic meters)	Per Capita Annual Water Withdrawal, Year of Data (cubic meters)	Sectoral Withdrawal as a Share of Total Water Resources (percent)		
						Agriculture	Domestic	Industry
Low-income	14,272	1,257	9	4,649	498	91	4	5
China and India	4,650	840	18	2,345	520	90	5	6
Other low-income	9,622	417	4	8,855	460	95	3	2
Middle-income	13,730	492	4	12,597	532	69	13	18
Lower-middle-income	6,483	290	4	10,259	550	71	11	18
Upper-middle-income	7,247	202	3	15,824	508	66	16	18
Low- and middle-income	28,002	1,749	6	6,732	507	85	7	8
Sub-Saharan Africa	3,713	55	1	7,488	140	88	8	3
East Asia and the Pacific	7,915	631	8	5,009	453	86	6	8
South Asia	4,895	569	12	4,236	652	94	2	3
Europe	574	110	19	2,865	589	45	14	42
Middle East and North Africa	276	202	73	1,071	1,003	89	6	5
Latin America and the Caribbean	10,579	173	2	24,390	460	72	16	11
Other economies	4,486	375	8	13,976	1,324	66	6	28
High-income	8,368	893	11	10,528	1,217	39	14	47
OECD members	8,365	889	11	10,781	1,230	39	14	47
Other	4	4	119	186	372	67	22	12
World	40,856	3,017	7	7,744	676	69	9	22

APPENDIX XV

MONEY AND INTEREST RATES IN VARIOUS COUNTRIES

	Money, Broadly Defined						Nominal Interest Rates of Banks (avg. annual %)			
	AVG. ANNUAL NOMINAL GROWTH RATE (%)		AVERAGE OUTSTANDING AS PERCENTAGE OF GDP			Avg. Annual Inflation (GDP Deflator)	DEPOSIT RATE		LENDING RATE	
	1970–80	1980–92	1970	1980	1992	1980–92	1980	1992	1980	1992
Low-income economies Excluding China and India										
1 Mozambique	. .	. .	. .	. .	. .	38.0	. .	. .	. .	. .
2 Ethiopia	14.4	12.8	14.0	25.3	65.2	2.8	. .	3.6	. .	8.0
3 Tanzania	22.6	. .	22.9	37.2	. .	25.3	4.0	. .	11.5	. .
4 Sierra Leone	19.9	58.1	12.6	20.6	12.4	60.8	9.2	54.7	11.0	62.8
5 Nepal	19.9	19.9	10.6	21.9	35.7	9.2	4.0	*8.5*	14.0	*14.4*
6 Uganda	28.1	. .	16.3	12.7	. .	. .	6.8	35.8	10.8	*34.4*
7 Bhutan	. .	30.7	. .	. .	22.7	8.7	. .	8.0	. .	17.0
8 Burundi	20.1	*9.9*	9.1	13.5	. .	4.5	2.5	. .	12.0	. .
9 Malawi	14.7	18.0	21.7	20.5	22.7	15.1	7.9	16.5	16.7	22.0
10 Bangladesh	. .	18.7	. .	18.4	31.1	9.1	8.3	10.5	11.3	15.0
11 Chad	15.2	7.6	9.4	20.0	20.1	0.9	5.5	7.5	11.0	16.3
12 Guinea-Bissau	. .	*59.6*	. .	. .	12.2	59.3	. .	39.3	. .	50.3
13 Madagascar	13.8	16.0	17.3	22.3	20.7	16.4	5.6	. .	9.5	. .
14 Lao PDR	. .	. .	. .	. .	. .	. .	7.2	*14.0*	4.8	*15.0*
15 Rwanda	21.5	8.3	10.7	13.6	17.3	3.6	6.3	7.7	13.5	16.7
16 Niger	23.9	4.5	5.2	13.3	19.6	1.7	6.2	7.8	14.5	16.8
17 Burkina Faso	21.5	10.3	9.3	15.9	21.2	3.5	6.2	7.8	14.5	16.8
18 India	17.3	*16.8*	23.9	36.2	*44.1*	8.5	. .	. .	16.5	18.
19 Kenya	19.8	15.8	31.2	36.8	44.6	9.3	5.8	*13.7*	10.6	*18.8*
20 Mali	18.5	8.1	13.8	17.9	20.7	3.7	6.2	7.8	14.5	16.8

	Money, Broadly Defined					Avg. Annual Inflation (GDP Deflator)	Nominal Interest Rates of Banks (avg. annual %)			
	Avg. annual nominal growth rate (%)		Average outstanding as percentage of GDP				Deposit rate		Lending rate	
	1970–80	1980–92	1970	1980	1992	1980–92	1980	1992	1980	1992
21 Nigeria	34.3	18.0	9.2	23.8	19.5	19.4	5.3	18.0	8.4	24.8
22 Nicaragua	18.2	. .	14.1	22.1	. .	656.2	7.5	. .	. .	. .
23 Togo	22.2	*5.9*	17.2	29.0	*35.2*	4.2	6.2	7.8	14.5	17.5
24 Benin	19.0	6.1	10.1	17.1	28.1	1.7	6.2	7.8	14.5	16.8
25 Central African Republic	16.0	3.8	16.0	18.9	16.6	4.6	5.5	7.5	10.5	16.3
26 Pakistan	17.1	13.7	41.2	38.7	39.0	7.1	. .	. .	. .	. .
27 Ghana	36.4	42.8	18.0	16.2	14.5	38.7	11.5	16.3	19.0	. .
28 China	. .	25.6	. .	25.5	66.6	6.5	5.4	. .	5.0	. .
29 Tajikistan	. .	. .	. .	. .	. .	. .	. .	. .	. .	. .
30 Guinea	. .	. .	. .	. .	. .	. .	. .	. .	. .	. .
31 Mauritania	21.5	11.4	9.5	21.3	25.5	8.3	*5.5*	5.0	12.0	10.0
32 Sri Lanka	23.1	15.3	22.0	35.3	35.2	11.0	14.5	18.3	19.0	13.0
33 Zimbabwe	. .	. .	. .	. .	36.0	14.4	3.5	*3.8*	17.5	*15.5*
34 Honduras	16.0	13.9	19.5	22.6	30.7	7.6	7.0	12.3	18.5	21.7
35 Lesotho	. .	16.4	. .	. .	35.8	13.2	. .	10.6	11.0	18.3
36 Egypt, Arab Rep.	26.0	21.7	33.5	52.2	91.4	13.2	8.3	*12.0*	13.3	*19.0*
37 Indonesia	35.9	26.3	7.8	13.2	42.6	8.4	6.0	20.4	. .	24.0
38 Myanmar	15.1	15.8	23.9	23.9	27.9	14.8	1.5	. .	8.0	. .
39 Somalia	24.6	. .	17.6	17.8	. .	*49.7*	4.5	. .	7.5	. .
40 Sudan	28.3	*34.9*	17.5	32.5	. .	42.8	6.0	. .	. .	. .
41 Yemen, Rep.	. .	*18.7*	. .	. .	. .	. .	9.3	. .	. .	. .
42 Zambia	10.7	. .	29.9	32.6	. .	*48.4*	7.0	48.5	9.5	54.6

	Money, Broadly Defined					Avg. Annual Inflation (GDP Deflator)	*Nominal Interest Rates of Banks (avg. annual %)*			
	AVG. ANNUAL NOMINAL GROWTH RATE (%)		AVERAGE OUTSTANDING AS PERCENTAGE OF GDP				DEPOSIT RATE		LENDING RATE	
	1970–80	1980–92	1970	1980	1992	1980–92	1980	1992	1980	1992
Middle-income economies										
Lower-middle-income										
43 Côte d'Ivoire	22.6	3.1	24.7	26.7	31.3	1.9	6.2	7.8	14.5	16.8
44 Bolivia	29.4	236.7	14.8	16.2	32.5	220.9	18.0	23.2	28.0	45.5
45 Azerbaijan	. .	. .	. .	. .	. .	. .	. .	. .	. .	. .
46 Philippines	19.2	17.0	29.9	26.4	34.3	14.1	12.3	14.3	14.0	19.5
47 Armenia	. .	. .	. .	. .	. .	. .	. .	. .	. .	. .
48 Senegal	19.6	5.5	14.0	26.6	22.8	5.2	6.2	7.8	14.5	16.8
49 Cameroon	22.5	6.0	13.5	18.3	26.0	3.5	7.5	8.0	13.0	16.3
50 Kyrgyz Republic	. .	. .	. .	. .	. .	. .	. .	. .	. .	. .
51 Georgia	. .	. .	. .	. .	. .	. .	. .	. .	. .	. .
52 Uzbekistan	. .	. .	. .	. .	. .	. .	. .	. .	. .	. .
53 Papua Nw Guinea	. .	8.3	. .	32.9	33.1	5.1	6.9	7.9	11.2	14.5
54 Peru	33.6	296.6	17.8	16.4	11.1	311.7	. .	59.7	. .	173.8
55 Guatemala	18.6	18.6	17.1	20.5	23.3	16.5	9.0	10.4	11.0	19.5
56 Congo	15.7	6.5	16.5	14.7	22.7	0.5	6.5	7.8	11.0	16.3
57 Morocco	18.7	*14.5*	31.1	42.4	. .	6.9	4.9	*8.5*	7.0	*9.0*

	Money, Broadly Defined					Avg. Annual Inflation (GDP Deflator)	*Nominal Interest Rates of Banks (avg. annual %)*			
	AVG. ANNUAL NOMINAL GROWTH RATE (%)		AVERAGE OUTSTANDING AS PERCENTAGE OF GDP				DEPOSIT RATE		LENDING RATE	
	1970–80	1980–92	1970	1980	1992	1980–92	1980	1992	1980	1992
58 Dominican Republic	18.3	29.0	17.9	22.0	24.7	25.2	. .	. .	. .	. .
59 Ecuador	24.2	*37.2*	20.0	20.2	*13.4*	39.5	. .	47.4	9.0	60.2
60 Jordan	24.3	12.7	. .	. .	126.7	. .	. .	3.3	. .	9.8
61 Romania	. .	13.8	. .	33.4	23.6	13.1	. .	. .	. .	. .
62 El Salvador	17.3	17.8	22.5	28.1	30.5	17.2	. .	11.5	. .	16.4
63 Turkmenistan	. .	. .	. .	. .	. .	. .	. .	. .	. .	. .
64 Moldova	. .	. .	. .	. .	. .	. .	. .	. .	. .	. .
65 Lithuania	. .	. .	. .	. .	. .	20.7	. .	. .	. .	. .
66 Bulgaria	. .	. .	. .	. .	. .	11.7	. .	54.5	. .	64.1
67 Colombia	32.7	. .	20.0	23.7	27.9	25.0	. .	26.7	. .	37.3
68 Jamaica	15.7	26.1	31.4	35.4	41.2	21.5	10.3	38.4	13.0	53.4
69 Paraguay	27.0	35.9	7.7	10.1	23.2	25.2	. .	20.1	. .	28.0
70 Namibia	. .	. .	. .	. .	. .	12.3	. .	11.4	. .	20.2
71 Kazakhstan	. .	. .	. .	. .	. .	. .	. .	. .	. .	. .
72 Tunisia	20.3	*15.5*	33.0	42.1	. .	7.2	2.5	*7.4*	7.3	*9.9*

Note: Figures in italics are for years other than those specified.

	Money, Broadly Defined						*Nominal Interest Rates of Banks (avg. annual %)*			
	AVG. ANNUAL NOMINAL GROWTH RATE (%)		AVERAGE OUTSTANDING AS PERCENTAGE OF GDP			Avg. Annual Inflation (GDP Deflator)	DEPOSIT RATE		LENDING RATE	
	1970–80	1980–92	1970	1980	1992	1980–92	1980	1992	1980	1992
73 Ukraine	. .	. .	. .	. .	. .	. .	. .	. .	. .	. .
74 Algeria	24.1	*14.3*	52.6	58.5	. .	11.4	. .	. .	. .	. .
75 Thailand	*17.9*	*19.2*	*23.6*	37.3	*71.5*	4.2	12.0	*12.3*	18.0	*25.0*
76 Poland	. .	62.3	. .	57.0	29.4	67.9	3.0	6.1	8.0	39.0
77 Latvia	. .	. .	. .	. .	. .	15.3	. .	. .	. .	. .
78 Slovak Republic	. .	. .	. .	. .	. .	. .	. .	. .	. .	. .
79 Costa Rica	30.6	25.8	18.9	38.8	38.1	22.5	. .	15.8	. .	28.5
80 Turkey	32.9	*52.7*	27.9	17.2	*21.6*	46.3	8.0	68.7	25.7	. .
81 Iran, Islamic Rep.	33.5	*17.3*	*26.1*	54.4	. .	16.2	. .	. .	. .	. .
82 Panama	. .	. .	. .	. .	. .	2.1	. .	. .	. .	. .
83 Czech Republic	. .	. .	. .	. .	. .	. .	. .	. .	. .	. .
84 Russian Federation	. .	. .	. .	. .	. .	. .	. .	. .	. .	. .
85 Chile	194.2	29.5	12.5	21.0	35.4	20.5	37.7	18.3	47.1	23.9
86 Albania	. .	. .	. .	. .	. .	. .	. .	. .	. .	. .
87 Mongolia	. .	. .	. .	. .	. .	. .	. .	. .	. .	. .
88 Syrian Arab Rep.	26.5	. .	34.8	40.9	. .	*15.5*	5.0	. .	. .	. .
Upper-middle-income										
89 South Africa	15.6	*16.6*	59.9	50.9	56.2	14.3	5.5	13.8	9.5	18.9
90 Mauritius	24.3	21.9	37.5	41.1	69.2	8.6	. .	10.1	. .	17.1
91 Estonia	. .	. .	. .	. .	. .	20.2	. .	. .	. .	. .
92 Brazil	52.7	. .	23.0	18.4	. .	370.2	115.0	1,560.2	. .	. .
93 Botswana	. .	25.3	. .	28.2	29.9	12.6	5.0	12.5	8.5	14.0

	Money, Broadly Defined					Avg. Annual Inflation (GDP Deflator)	Nominal Interest Rates of Banks (avg. annual %)			
	AVG. ANNUAL NOMINAL GROWTH RATE (%)		AVERAGE OUTSTANDING AS PERCENTAGE OF GDP				DEPOSIT RATE		LENDING RATE	
	1970–80	1980–92	1970	1980	1992	1980–92	1980	1992	1980	1992
94 Malaysia	25.2	*12.6*	34.4	69.8	. .	2.0	6.2	*7.2*	7.8	*8.1*
95 Venezuela	26.4	21.7	24.1	43.0	34.9	22.7	. .	35.4	. .	33.9
96 Belarus	. .	. .	. .	. .	. .	. .	. .	. .	. .	. .
97 Hungary	. .	. .	. .	. .	. .	11.7	3.0	*23.0*	9.0	*30.0*
98 Uruguay	78.4	70.4	20.6	31.2	39.3	66.2	50.3	54.5	66.6	117.8
99 Mexico	26.6	60.3	26.9	27.5	28.2	62.4	20.6	15.7	28.1	. .
100 Trinidad and Tobago	27.1	5.6	28.2	30.5	51.6	3.9	. .	7.0	10.0	15.3
101 Gabon	31.3	4.5	14.5	15.2	18.7	2.3	7.5	8.8	12.5	12.5
102 Argentina	142.8	377.1	24.1	19.0	11.1	402.3	79.6	16.8	86.9	15.1
103 Oman	*29.4*	10.3	. .	13.8	28.3	−2.5	. .	6.3	. .	9.2
104 Slovenia	. .	. .	. .	. .	. .	. .	. .	. .	. .	. .
105 Puerto Rico	. .	. .	. .	. .	. .	3.3	. .	. .	. .	. .
106 Korea, Rep.	30.4	21.6	32.1	31.7	57.9	5.9	19.5	10.0	18.0	10.0
107 Greece	23.9	*22.3*	42.9	61.6	*79.3*	17.7	14.5	19.9	21.3	28.7
108 Portugal	20.2	18.7	87.6	80.8	80.8	*17.4*	19.0	14.6	18.8	20.4
109 Saudi Arabia	43.7	*7.6*	17.6	18.6	. .	−1.9	. .	. .	. .	. .

	Money, Broadly Defined					Avg. Annual Inflation (GDP Deflator)	*Nominal Interest Rates of Banks (avg. annual %)*			
	AVG. ANNUAL NOMINAL GROWTH RATE (%)		AVERAGE OUTSTANDING AS PERCENTAGE OF GDP				DEPOSIT RATE		LENDING RATE	
	1970–80	1980–92	1970	1980	1992	1980–92	1980	1992	1980	1992
Low- and middle-income										
Sub-Saharan Africa										
East Asia and Pacific										
South Asia										
Europe and Central Asia										
Middle East and N. Africa										
Latin America and Caribbean										
Severely indebted										
High-income economies										
110 Ireland	19.1	6.8	64.0	58.1	46.9	5.3	12.0	5.4	16.0	*10.6*
111 New Zealand	15.1	. .	51.4	50.9	. .	9.4	. .	6.6	12.6	11.4
112 Israel	36.2	87.6	36.5	14.7	57.5	78.9	. .	11.3	176.9	19.9
113 Spain	20.1	12.5	69.5	75.4	76.8	8.7	13.1	10.4	16.9	14.2
114 Hong Kong	. .	. .	. .	69.5	. .	7.8	. .	. .	. .	. .
115 Singapore	17.1	13.6	66.2	74.4	129.0	2.0	9.4	2.9	11.7	6.0
116 Australia	16.8	11.5	43.6	46.5	57.9	6.4	8.6	*10.4*	10.6	12.0
117 United Kingdom	15.2	. .	49.2	46.0	. .	5.7	14.1	7.3	16.2	9.4
118 Italy	20.4	10.5	79.3	83.1	72.9	9.1	12.7	7.1	19.0	15.8
119 Netherlands	14.6	. .	54.3	77.7	. .	1.7	6.0	3.2	13.5	12.8

	Money, Broadly Defined					Avg. Annual Inflation (GDP Deflator)	Nominal Interest Rates of Banks (avg. annual %)			
	AVG. ANNUAL NOMINAL GROWTH RATE (%)		AVERAGE OUTSTANDING AS PERCENTAGE OF GDP				DEPOSIT RATE		LENDING RATE	
	1970–80	1980–92	1970	1980	1992	1980–92	1980	1992	1980	1992
120 Canada	17.5	8.3	48.4	65.0	77.9	4.1	12.9	6.7	14.3	7.5
121 Belgium	10.8	*7.0*	56.7	57.0	. .	4.1	7.7	6.3	. .	13.0
122 Finland	15.4	12.2	39.8	39.5	61.9	6.0	. .	7.5	9.8	12.1
123 United Arab Emirates	. .	8.6	. .	19.0	52.6	*0.8*	9.5	. .	12.1	. .
124 France	15.6	*9.9*	57.8	69.7	. .	5.4	6.3	. .	18.7	. .
125 Austria	13.7	7.4	54.0	72.6	88.0	3.6	5.0	3.7	. .	. .
126 Germany[a]	9.4	6.6	52.8	60.7	68.3	2.7	8.0	8.0	12.0	13.6
127 United States	10.0	7.6	60.4	58.3	65.4	3.9	. .	. .	15.3	6.3
128 Norway	12.8	10.1	54.6	51.6	66.2	4.9	5.0	10.7	12.6	14.3
129 Denmark	12.4	10.1	44.8	42.6	58.6	4.9	10.8	7.5	17.2	11.6
130 Sweden	11.5	7.2	55.2	53.9	47.0	7.2	11.3	7.8	15.1	15.2
131 Japan	16.0	8.5	94.7	134.1	184.6	1.5	5.5	2.7	8.4	6.2
132 Switzerland	5.4	6.4	109.8	107.4	112.4	3.8	. .	5.5	. .	7.8
World										

[a]Data refer to the Federal Republic of Germany before unification.

APPENDIX XVI

EDUCATION PROFILE OF COUNTRIES

	Percentage of Age Group Enrolled in Education												Primary Pupil/Teacher Ratio	
	PRIMARY				SECONDARY									
	Total		Female		Total		Female		Tertiary		Primary Net Enrollment (%)			
	1970	1991	1970	1991	1970	1991	1970	1991	1970	1991	1975	1991	1970	1991
Low-income economies	**74**	**101**	**. .**	**93**	**21**	**41**	**. .**	**35**	**. .**	**3**	**. .**	**. .**	**36**	**38**
Excluding China & India	**55**	**79**	**44**	**71**	**13**	**28**	**8**	**25**	**3**	**5**	**. .**	**74**	**39**	**38**
1 Mozambique	47	63	. .	53	5	8	. .	5	0	. .	. .	*41*	69	55
2 Ethiopia	16	25	10	21	4	12	2	11	0	1	. .	. .	48	30
3 Tanzania	34	69	27	68	3	5	2	4	*0*	*0*	. .	*47*	47	36
4 Sierra Leone	34	*48*	27	*39*	8	*16*	5	*12*	1	*1*	. .	. .	32	*34*
5 Nepal	26	. .	8	. .	10	. .	3	. .	3	7	. .	. .	22	36
6 Uganda	38	71	30	63	4	13	2	35	1	*1*	. .	. .	34	. .
7 Bhutan	6	. .	1	. .	1	. .	0	. .	0	. .	. .	. .	21	. .
8 Burundi	30	70	20	63	2	6	1	4	1	1	. .	. .	37	66
9 Malawi	. .	*66*	. .	*60*	. .	*4*	. .	*3*	1	*1*	. .	*54*	43	*64*
10 Bangladesh	54	*77*	35	*71*	. .	*19*	. .	12	3	*4*	. .	*65*	46	*63*
11 Chad	35	65	17	41	2	*7*	0	*3*	. .	. .	. .	. .	65	64
12 Guinea-Bissau	39	. .	23	. .	8	. .	6	. .	0	. .	59	. .	45	. .
13 Madagascar	90	*92*	82	*91*	12	*19*	9	*18*	3	*3*	. .	*64*	65	*40*
14 Lao PDR	53	98	40	84	3	22	2	17	1	*1*	. .	*69*	36	28
15 Rwanda	68	*71*	60	*70*	2	*8*	1	*7*	0	*1*	. .	*65*	60	58
16 Niger	14	*29*	10	*21*	1	*6*	1	*4*	0	*1*	. .	*25*	39	*42*
17 Burkina Faso	13	30	10	24	1	8	1	*5*	0	*1*	. .	*29*	44	58
18 India	73	*98*	56	*84*	26	*44*	15	*32*	. .	. .	. .	. .	41	60
19 Kenya	58	*95*	48	*93*	9	*29*	5	*25*	1	*2*	88	. .	34	*31*
20 Mali	22	25	15	19	5	7	2	5	0	*1*	. .	*19*	40	47

		Percentage of Age Group Enrolled in Education												Primary Pupil/ Teacher Ratio	
		PRIMARY				SECONDARY				Tertiary		Primary Net Enrollment (%)			
		Total		Female		Total		Female							
		1970	1991	1970	1991	1970	1991	1970	1991	1970	1991	1975	1991	1970	1991
21	Nigeria	37	71	27	62	4	20	3	17	2	*4*	. .	. .	34	39
22	Nicaragua	80	101	81	104	18	44	17	46	14	*10*	65	*75*	37	36
23	Togo	71	*111*	44	*87*	7	*23*	3	*12*	2	*3*	. .	. .	58	*59*
24	Benin	36	66	22	*39*	5	12	3	7	2	*3*	. .	. .	41	*35*
25	Central African Republic	64	*68*	41	*52*	4	*12*	2	*7*	1	*2*	. .	*55*	64	*90*
26	Pakistan	40	*46*	22	*31*	13	*21*	5	*13*	*4*	*3*	. .	. .	41	*41*
27	Ghana	64	*77*	54	*69*	14	*38*	8	*29*	2	*2*	. .	. .	30	*29*
28	China	89	123	. .	118	24	51	. .	45	1	2	. .	*100*	29	22
29	Tajikistan	. .	. .	. .	. .	. .	. .	. .	. .	. .	. .	. .	. .	. .	. .
30	Guinea	33	37	21	*24*	13	*10*	5	*5*	5	. .	. .	*26*	44	49
31	Mauritania	14	55	8	48	2	14	0	10	. .	3	. .	. .	24	47
32	Sri Lanka	99	108	94	106	47	74	48	77	3	*5*	. .	. .	. .	12
33	Zimbabwe	74	*117*	66	120	7	52	6	45	1	5	. .	. .	. .	39
34	Honduras	87	105	87	107	14	19	13	34	8	*9*	. .	. .	35	38
35	Lesotho	87	107	101	*116*	7	25	7	*30*	2	3	. .	*70*	46	54
36	Egypt, Arab Rep.	72	101	57	*93*	35	80	23	*73*	18	*19*	. .	. .	38	24
37	Indonesia	80	*116*	73	*114*	16	*45*	11	*41*	*4*	10	72	*98*	29	*23*
38	*Myanmar*	83	*102*	78	. .	21	*20*	16	. .	*5*	. .	. .	. .	47	*35*
39	*Somalia*	11	. .	5	. .	5	. .	2	. .	. .	. .	16	. .	33	. .
40	*Sudan*	38	*50*	29	*43*	7	22	4	*20*	2	*3*	. .	. .	47	34

	Percentage of Age Group Enrolled in Education												Primary Pupil/ Teacher Ratio	
	PRIMARY				SECONDARY				Tertiary		Primary Net Enrollment (%)			
	Total		Female		Total		Female							
	1970	1991	1970	1991	1970	1991	1970	1991	1970	1991	1975	1991	1970	1991
41 *Yemen. Rep.*	22	*76*	7	*37*	3	*31*	0	. .	. .	. .	. .	. .	51	*37*
42 *Zambia*	90	*92*	80	. .	13	. .	8	. .	2	2	. .	. .	47	. .
Middle-income economies	**93**	**104**	**87**	**99**	**32**	**55**	**26**	**56**	**13**	**18**	**. .**	**90**	**34**	**25**
Lower-middle-income	**. .**	**. .**	**. .**	**. .**	**. .**	**. .**	**. .**	**. .**	**. .**	**. .**	**. .**	**. .**	**. .**	**. .**
43 Côte d'Ivoire	58	69	45	58	9	24	4	16	3	. .	. .	. .	45	37
44 Bolivia	76	*85*	62	*81*	24	*34*	20	*31*	13	23	73	*79*	27	*25*
45 Azerbaijan	. .	. .	. .	. .	. .	. .	. .	. .	. .	. .	. .	. .	. .	. .
46 Philippines	108	110	. .	*111*	46	74	. .	*75*	28	28	95	*99*	29	33
47 Armenia	. .	. .	. .	. .	. .	. .	. .	. .	. .	. .	. .	. .	. .	. .
48 Senegal	41	*59*	32	*49*	10	*16*	6	*11*	3	*3*	. .	*48*	45	*58*
49 Cameroon	89	*101*	75	*93*	7	*28*	4	*23*	2	*3*	69	*75*	48	*51*
50 Kyrgyz Republic	. .	. .	. .	. .	. .	. .	. .	. .	. .	. .	. .	. .	. .	. .
51 Georgia	. .	. .	. .	. .	. .	. .	. .	. .	. .	. .	. .	. .	. .	. .
52 Uzbekistan	. .	. .	. .	. .	. .	. .	. .	. .	. .	. .	. .	. .	. .	. .
53 Papua New Guinea	52	*71*	39	*65*	8	*12*	4	*10*	2	. .	. .	*73*	30	31
54 Peru	107	*126*	99	. .	31	*70*	27	. .	19	*36*	. .	. .	35	*28*
55 Guatemala	57	79	51	73	8	28	8	. .	8	. .	53	. .	36	34
56 Congo	. .	. .	. .	. .	. .	. .	. .	. .	5	6	. .	. .	62	*66*
57 Monocco	52	66	36	54	13	28	7	29	6	*10*	47	. .	34	27

		Percentage of Age Group Enrolled in Education												Primary Pupil/ Teacher Ratio	
		PRIMARY				SECONDARY									
		Total		Female		Total		Female		Tertiary		Primary Net Enrollment (%)			
		1970	1991	1970	1991	1970	1991	1970	1991	1970	1991	1975	1991	1970	1991
58	Dominican Republic	100	. .	100	. .	21	. .	. .	. .	. .	. .	. .	. .	55	*47*
59	Ecuador	97	. .	95	. .	22	. .	23	. .	37	*20*	78	. .	38	. .
60	Jordan	. .	*97*	. .	*98*	. .	*91*	. .	*62*	27	*25*	. .	. .	39	24
61	Romania	112	90	113	90	44	80	38	80	11	*9*	. .	. .	21	17
62	El Salvador	85	76	83	77	22	25	21	27	4	*16*	. .	*70*	36	44
63	Turkmenistan	. .	. .	. .	. .	. .	. .	. .	. .	. .	. .	. .	. .	. .	. .
64	Moldova	. .	. .	. .	. .	. .	. .	. .	. .	. .	. .	. .	. .	. .	. .
65	Lithuania	. .	. .	. .	. .	. .	. .	. .	. .	. .	. .	. .	. .	. .	. .
66	Bulgaria	101	92	100	91	79	71	. .	73	16	30	96	*85*	22	15
67	Colombia	108	111	110	112	25	55	24	60	10	*14*	. .	*73*	38	30
68	Jamaica	119	*106*	119	*108*	46	*62*	45	*66*	7	*6*	90	*99*	47	*37*
69	Paraguay	109	109	103	108	17	30	17	31	9	*8*	83	*95*	32	25
70	Namibia	. .	*119*	. .	*126*	. .	*41*	. .	*47*	. .	*3*	. .	. .	. .	. .
71	Kazakhstan	. .	. .	. .	. .	. .	. .	. .	. .	. .	. .	. .	. .	. .	. .
72	Tunisia	100	117	79	110	23	46	13	42	5	9	. .	*95*	47	26
73	Ukraine	. .	. .	. .	. .	. .	. .	. .	. .	. .	. .	. .	. .	15	*8*
74	Algeria	76	*95*	58	*88*	11	*60*	6	*53*	6	*12*	77	*88*	40	28
75	Thailand	83	113	79	*88*	17	*33*	15	*32*	13	*16*	. .	. .	35	*18*
76	Poland	101	98	99	97	62	83	65	86	18	22	96	*97*	23	17
77	Latvia	. .	. .	. .	. .	. .	. .	. .	. .	. .	. .	. .	. .	. .	. .

	Percentage of Age Group Enrolled in Education												Primary Pupil/ Teacher Ratio	
	PRIMARY				SECONDARY						Primary Net Enrollment (%)			
	Total		Female		Total		Female		Tertiary					
	1970	1991	1970	1991	1970	1991	1970	1991	1970	1991	1975	1991	1970	1991
78 Slovak Republic	. .	100	. .	. .	. .	97	. .	. .	. .	*27*	. .	. .	. .	*19*
79 Costa Rica	110	103	109	102	28	43	29	45	23	28	92	*87*	30	32
80 Turkey	110	*110*	94	110	27	51	15	40	6	15	. .	*99*	38	29
81 Iran, Islamic Rep.	72	112	52	105	27	57	18	49	*4*	12	. .	*94*	32	31
82 Panama	99	*106*	97	*105*	38	*60*	40	*62*	22	24	87	*92*	27	*20*
83 Czech Republic	. .	. .	. .	. .	. .	. .	. .	. .	. .	. .	. .	. .	. .	. .
84 Russian Federation	. .	. .	. .	. .	. .	. .	. .	. .	. .	. .	. .	. .	. .	. .
85 Chile	107	98	107	97	39	72	42	75	13	23	94	*86*	50	25
86 *Albania*	106	*101*	102	*101*	35	*79*	27	*74*	5	*7*	. .	. .	26	*19*
87 *Mongolia*	113	89	. .	*100*	87	77	. .	. .	. .	*15*	. .	. .	30	25
88 *Syrian Arab Rep.*	78	109	59	103	38	50	21	43	18	19	87	*98*	37	25
Upper-middle-income	**94**	**105**	**92**	**105**	**32**	**54**	**29**	**64**	**14**	**19**	**80**	**90**	**34**	**24**
89 South Africa	99	. .	99	. .	18	. .	17	. .	. .	. .	. .	. .	34	. .
90 Mauritius	94	106	93	108	30	54	25	56	1	*2*	82	*92*	32	21
91 Estonia	. .	. .	. .	. .	. .	. .	. .	. .	. .	. .	. .	. .	. .	. .
92 Brazil	*82*	106	*82*	. .	26	39	26	. .	12	12	71	*88*	28	23
93 Botswana	65	119	67	121	7	54	6	57	1	3	58	*91*	36	30
94 Malaysia	87	93	84	93	34	58	28	59	4	*7*	. .	. .	31	20
95 Venezuela	94	99	94	100	33	34	34	40	21	*30*	81	*61*	35	23
96 Belarus	. .	. .	. .	. .	. .	. .	. .	. .	. .	. .	. .	. .	. .	. .
97 Hungary	97	89	97	89	63	81	55	81	13	15	. .	*90*	18	12
98 Uruguay	112	108	109	107	59	84	64	. .	18	32	. .	. .	29	22

	Percentage of Age Group Enrolled in Education												Primary Pupil/Teacher Ratio	
	PRIMARY				SECONDARY									
	Total		Female		Total		Female		Tertiary		Primary Net Enrollment (%)			
	1970	1991	1970	1991	1970	1991	1970	1991	1970	1991	1975	1991	1970	1991
99 Mexico	104	114	101	112	22	55	17	55	14	*15*	. .	*98*	46	30
100 Trinidad and Tobago	106	*96*	107	*96*	42	*81*	44	*82*	5	*7*	87	*91*	34	*26*
101 Gabon	85	. .	81	. .	8	. .	5	. .	. .	3	. .	. .	46	44
102 Argentina	105	107	106	*114*	44	. .	47	. .	22	*43*	96	. .	19	18
103 Oman	3	100	1	96	. .	57	. .	53	0	6	32	*84*	18	27
104 Slovenia	. .	. .	. .	. .	. .	. .	. .	. .	. .	. .	. .	. .	. .	. .
105 Puerto Rico	117	. .	. .	. .	71	. .	. .	. .	48	. .	. .	. .	30	. .
106 Korea, Rep.	103	107	103	109	42	88	32	88	16	40	99	*100*	57	34
107 Greece	107	*97*	106	*98*	63	*98*	55	*94*	17	*25*	97	. .	31	*20*
108 Portugal	98	122	96	*115*	57	*68*	51	*74*	11	*23*	91	*99*	34	*14*
109 Saudi Arabia	45	*77*	29	*72*	12	*46*	5	*41*	7	*13*	42	*62*	24	*16*
Low- and middle-income	**79**	**102**	**63**	**94**	**24**	**45**	**17**	**39**	**6**	**8**	**. .**	**92**	**35**	**35**
Sub-Saharan Africa	50	66	41	58	7	18	5	16	1	2	. .	. .	42	41
East Asia & Pacific	88	119	. .	115	24	50	. .	47	4	5	. .	100	30	24
South Asia	67	89	50	76	25	39	14	29	. .	. .	. .	. .	42	57
Europe and Central Asia	. .	. .	. .	. .	. .	. .	. .	. .	. .	. .	. .	. .	. .	. .
Middle East & N. Africa	68	98	50	89	24	56	15	51	10	15	. .	89	35	27
Latin America & Caribbean	95	106	94	105	28	47	26	54	15	18	. .	87	34	26
Severely indebted	**90**	**103**	**85**	**97**	**30**	**50**	**27**	**54**	**14**	**17**	**. .**	**91**	**32**	**25**
High-income economies	**106**	**104**	**106**	**103**	**73**	**93**	**71**	**95**	**36**	**50**	**88**	**99**	**26**	**17**
110 Ireland	106	*103*	106	*103*	74	*101*	77	*105*	20	*34*	91	*88*	24	*27*
111 New Zealand	110	*104*	109	*103*	77	*84*	76	*85*	29	45	100	*100*	21	*19*
112 Israel	96	*95*	95	*96*	57	*85*	60	*89*	29	*34*	. .	. .	17	17
113 Spain	123	*109*	125	*108*	56	*108*	48	*113*	24	*36*	100	. .	34	*21*
114 Hong Kong	117	*108*	115	. .	36	. .	31	. .	11	*18*	92	. .	33	*27*

		Percentage of Age Group Enrolled in Education												Primary Pupil/ Teacher Ratio	
		PRIMARY				SECONDARY						Primary Net Enrollment (%)			
		Total		Female		Total		Female		Tertiary					
		1970	1991	1970	1991	1970	1991	1970	1991	1970	1991	1975	1991	1970	1991
115	Singapore	105	*108*	101	*107*	46	*70*	45	*71*	8	. .	100	*100*	30	*26*
116	Australia	115	*107*	115	107	82	82	80	83	25	39	98	*97*	28	17
117	United Kingdom	104	*104*	104	*105*	73	*86*	73	*88*	20	*28*	97	*100*	23	*20*
118	Italy	110	94	109	94	61	76	55	76	28	32	97	. .	22	12
119	Netherlands	102	*102*	102	*103*	75	97	69	96	30	*38*	92	*100*	30	*17*
120	Canada	101	*107*	100	*106*	65	*104*	65	*104*	42	99	. .	*96*	23	*15*
121	Belgium	103	*99*	104	100	81	102	80	103	26	*38*	. .	*99*	20	10
122	Finland	82	99	79	99	102	121	106	133	32	51	. .	. .	22	*18*
123	United Arab Emirates	93	115	71	114	22	69	9	73	2	11	. .	*100*	27	18
124	France	117	107	117	106	74	101	77	104	26	43	98	*100*	26	12
125	Austria	104	*103*	103	*102*	72	104	73	100	23	35	89	. .	21	11
126	Germany	. .	*107*	. .	*107*	. .	. .	. .	*103*	27	*36*	. .	. .	. .	17
127	United States	. .	*104*	. .	*104*	. .	*90*	. .	*90*	56	*76*	72	*99*	27	. .
128	Norway	89	100	94	100	83	103	83	104	26	45	100	*98*	20	*6*
129	Denmark	96	*96*	97	*96*	78	*108*	75	*110*	29	*36*	. .	. .	9	*11*
130	Sweden	94	100	95	100	86	91	85	93	31	34	100	*100*	20	6
131	Japan	99	102	99	102	86	*97*	86	*98*	31	*31*	99	*100*	26	21
132	Switzerland	. .	103	. .	104	. .	91	. .	88	18	29	. .	. .	. .	. .
World		**83**	**102**	**71**	**96**	**31**	**52**	**28**	**49**	**12**	**17**	**. .**	**94**	**33**	**33**

Note: Figures in italics are for years other than those specified.

Source: World Development Report, *Workers in an Integrating World,* World Bank Report, 1995.

APPENDIX XVII

BASIC SOCIOECONOMIC INDICATORS OF VARIOUS COUNTRIES

	Population (millions) mid-1992	Area (thousands of sq. km)	GNP per Capita[a]		Avg. Annual Rate of Inflation (%)		Life Expectancy at Birth (years) 1992	Adult Illiteracy (%)	
			Dollars 1992	Avg. Ann. Growth (%), 1980–92	1970–80	1980–92		Female 1990	Total 1990
Low-income economies	3,191.3	38,929	390	3.9	. .	12.2	62	52	40
Excluding China and India	1,145.6	26,080	370	1.2	15.7	22.1	56	56	45
1 Mozambique	16.5	802	60	−3.6	. .	38.0	44	79	67
2 Ethiopia	54.8	1,222	110	−1.9	4.3	2.8	49	. .	. .
3 Tanzania[a]	25.9	945	110	0.0	14.1	25.3	51	. .	. .
4 Sierra Leone	4.4	72	160	−1.4	12.5	60.8	43	89	79
5 Nepal	19.9	141	170	2.0	8.5	9.2	54	87	74
6 Uganda	17.5	236	170	. .	. .	. .	43	65	52
7 Bhutan	1.5	47	180	6.3	. .	8.7	48	75	62
8 Burundi	5.8	28	210	1.3	10.7	4.5	48	60	50
9 Malawi	9.1	118	210	−0.1	8.8	15.1	44	. .	. .
10 Bangladesh	114.4	144	220	1.8	20.8	9.1	55	78	65
11 Chad	6.0	1,284	220	3.4	7.7	0.9	47	82	70
12 Guinea-Bissau	1.0	36	220	1.6	5.7	59.3	39	76	64
13 Madagascar	12.4	587	230	−2.4	9.9	16.4	51	27	20
14 Lao PDR	4.4	237	250	. .	. .	. .	51	. .	. .
15 Rwanda	7.3	26	250	−0.6	15.1	3.6	46	63	50
16 Niger	8.2	1,267	280	−4.3	10.9	1.7	46	83	72
17 Burkina Faso	9.5	274	300	1.0	8.6	3.5	48	91	82
18 India	883.6	3,288	310	3.1	8.4	8.5	61	66	52
19 Kenya	25.7	580	310	0.2	10.1	9.3	59	42	31
20 Mali	9.0	1,240	310	−2.7	9.7	3.7	48	76	68
21 Nigeria	101.9	924	320	−0.4	15.2	19.4	52	61	49
22 Nicaragua	3.9	130	340	−5.3	12.8	656.2	67	. .	. .
23 Togo	3.9	57	390	−1.8	8.9	4.2	55	69	57
24 Benin	5.0	113	410	−0.7	10.3	1.7	51	84	77
25 Central African Republic	3.2	623	410	−1.5	12.1	4.6	47	75	62
26 Pakistan	119.3	796	420	3.1	13.4	7.1	59	79	65
27 Ghana	15.8	239	450	−0.1	35.2	38.7	56	49	40
28 China	1,162.2	9,561	470	7.6	. .	6.5	69	38	27
29 Tajikistan[b]	5.6	143	490	. .	. .	. .	69	. .	. .
30 Guinea	6.1	246	510	. .	. .	. .	44	87	76

	Population (millions) mid-1992	Area (thousands of sq. km)	GNP per Capita[a]		Avg. Annual Rate of Inflation (%)		Life Expectancy at Birth (years) 1992	Adult Illiteracy (%)	
			Dollars 1992	Avg. Ann. Growth (%), 1980–92	1970–80	1980–92		Female 1990	Total 1990
31 Mauritania	2.1	1,026	530	−0.8	9.9	8.3	48	79	66
32 Sri Lanka	17.4	66	540	2.6	12.3	11.0	72	17	12
33 Zimbabwe	10.4	391	570	−0.9	9.4	14.4	60	40	33
34 Honduras	5.4	112	580	−0.3	8.1	7.6	66	29	27
35 Lesotho	1.9	30	590	−0.5	9.7	13.2	60	. .	. .
36 Egypt. Arab Rep.	54.7	1,001	640	1.8	9.6	13.2	62	66	52
37 Indonesia	184.3	1,905	670	4.0	21.5	8.4	60	32	23
38 Myanmar	43.7	677	. .	. .	11.4	14.8	60	28	19
39 Somalia	8.3	638	. .	. .	15.2	49.7	49	86	76
40 Sudan	26.5	2,506	. .	. .	14.5	42.8	52	88	73
41 Yemen, Rep.	13.0	528	. .	. .	. .	. .	53	74	62
42 Zambia	8.3	753	. .	. .	7.6	48.4	48	35	27
Middle-income economies	1,418.7	62,740	2,490	−0.1	31.0	105.2	68	. .	. .
Lower-middle-income	941.0	40,903	. .	. .	23.8	40.7	67	. .	. .
43 Côte d'Ivoire	12.9	322	670[c]	−4.7	13.0	1.9	56	60	46
44 Bolivia	7.5	1,099	680	−1.5	21.0	220.9	60	29	23
45 Azerbaijan[b]	7.4	87	740	. .	. .	. .	71	. .	. .
46 Philippines	64.3	300	770	−1.0	13.3	14.1	65	11	10
47 Armenia[b]	3.7	30	780	. .	. .	. .	70	. .	. .
48 Senegal	7.8	197	780	0.1	8.5	5.2	49	75	62
49 Cameroon	12.2	475	820	−1.5	9.8	3.5	56	57	46
50 Kyrgyz Republic[b]	4.5	199	820	. .	. .	. .	66	. .	. .
51 Georgia[b]	5.5	70	850	. .	. .	. .	72	. .	. .
52 Uzbekistan[b]	21.5	447	850	. .	. .	. .	69	. .	. .
53 Papua New Guinea	4.1	463	950	0.0	9.1	5.1	56	62	48
54 Peru	22.4	1,285	950	−2.8	30.1	311.7	65	21	15
55 Guatemala	9.7	109	980	−1.5	10.5	16.5	65	53	45
56 Congo	2.4	342	1,030	−0.8	8.4	0.5	51	56	43
57 Morocco	26.2	447	1,030	1.4	8.3	6.9	63	62	51

	Population (millions) mid-1992	Area (thousands of sq. km)	GNP per Capita[a]		Avg. Annual Rate of Inflation (%)		Life Expectancy at Birth (years) 1992	Adult Illiteracy (%)	
			Dollars 1992	Avg. Ann. Growth (%), 1980–92	1970–80	1980–92		Female 1990	Total 1990
58 Dominican Republic	7.3	49	1,050	−0.5	9.1	25.2	68	18	17
59 Ecuador	11.0	284	1,070	−0.3	13.8	39.5	67	16	14
60 Jordan[d]	3.9	89	1,120	*−5.4*	. .	*5.4*	70	30	20
61 Romania	22.7	238	1,130	−1.1	. .	13.1	70	. .	. .
62 El Salvador	5.4	21	1,170	0.0	10.7	17.2	66	30	27
63 Turkmenistan[b]	3.9	488	1,230	. .	. .	. .	66	. .	. .
64 Moldova[b]	4.4	34	1,300	. .	. .	. .	68	. .	. .
65 Lithuania[b]	3.8	65	1,310	−1.0	. .	20.7	71	. .	. .
66 Bulgaria	8.5	111	1,330	1.2	. .	11.7	71	. .	. .
67 Colombia	33.4	1,139	1,330	1.4	22.3	25.0	69	14	13
68 Jamaica	2.4	11	1,340	0.2	17.3	21.5	74	1	2
69 Paraguay	4.5	407	1,380	−0.7	12.7	25.2	67	12	10
70 Namibia	1.5	824	1,610	−1.0	. .	12.3	59	. .	. .
71 Kazakhstan[b]	17.0	2,717	1,680	. .	. .	. .	68	. .	. .
72 Tunisia	8.4	164	1,720	1.3	8.7	7.2	68	44	35
73 Ukraine[b]	52.1	604	1,820	. .	. .	. .	70	. .	. .
74 Algeria	26.3	2,382	1,840	−0.5	14.5	11.4	67	55	43
75 Thailand	58.0	513	1,840	6.0	9.2	4.2	69	10	7
76 Poland	38.4	313	1,910	0.1	. .	67.9	70	. .	. .
77 Latvia[b]	2.6	65	1,930	0.2	. .	15.3	69	. .	. .

	Population (millions) mid-1992	Area (thousands of sq. km)	GNP per Capita[a]		Avg. Annual Rate of Inflation (%)		Life Expectancy at Birth (years) 1992	Adult Illiteracy (%)	
			Dollars 1992	Avg. Ann. Growth (%), 1980–92	1970–80	1980–92		Female 1990	Total 1990
78 Slovak Republic	5.3	49	1,930	. .	. .	. .	71	. .	. .
79 Costa Rica	3.2	51	1,960	0.8	15.3	22.5	76	7	7
80 Turkey	58.5	779	1,980	2.9	29.4	46.3	67	29	19
81 Iran, Islamic Rep.	59.6	1,648	2,200	−1.4	. .	16.2	65	57	46
82 Panama	2.5	77	2,420	−1.2	7.7	2.1	73	12	12
83 Czech Republic	10.3	79	2,450	. .	. .	. .	72	. .	. .
84 Russian Federation[b]	149.0	17,075	2,510	. .	. .	. .	69	. .	. .
85 Chile	13.6	757	2,730[c]	3.7	187.1	20.5	72	7	7
86 Albania	3.4	29	. .	. .	. .	. .	73	. .	. .
87 Mongolia	2.3	1,567	. .	. .	. .	. .	64	. .	. .
88 Syrian Arab Rep.	13.0	185	. .	. .	11.8	15.5	67	49	36
Upper-middle-income	477.7	21,837	4,020	0.8	34.5	154.8	69	18	15
89 South Africa	39.8	1,221	2,670[c]	0.1	13.0	14.3	63	. .	. .
90 Mauritius	1.1	2	2,700	5.6	15.3	8.6	70	. .	. .
91 Estonia[b]	1.6	45	2,760	−2.3	. .	20.2	70	. .	. .
92 Brazil	153.9	8,512	2,770	0.4	38.6	370.2	66	20	19
93 Botswana	1.4	582	2,790	6.1	11.6	12.6	68	35	26
94 Malaysia	18.6	330	2,790	3.2	7.3	2.0	71	30	22
95 Venezuela	20.2	912	2,910	−0.8	14.0	22.7	70	17	8
96 Belarus[b]	10.3	208	2,930	. .	. .	. .	71	. .	. .
97 Hungary	10.3	93	2,970	0.2	2.8	11.7	69	. .	. .
98 Uruguay	3.1	177	3,340	−1.0	63.9	66.2	72	4	4
99 Mexico	85.0	1,958	3,470	−0.2	18.1	62.4	70	15	13
100 Trinidad and Tobago	1.3	5	3,940	−2.6	18.5	3.9	71	. .	. .
101 Gabon	1.2	268	4,450	−3.7	17.5	2.3	54	52	39
102 Argentina	33.1	2,767	6,050	−0.9	134.2	402.3	71	5	5
103 Oman	1.6	212	6,480	4.1	28.0	−2.5	70	. .	. .

	Population (millions) mid-1992	Area (thousands of sq. km)	GNP per Capita[a] Dollars 1992	GNP per Capita[a] Avg. Ann. Growth (%), 1980–92	Avg. Annual Rate of Inflation (%) 1970–80	Avg. Annual Rate of Inflation (%) 1980–92	Life Expectancy at Birth (years) 1992	Adult Illiteracy (%) Female 1990	Adult Illiteracy (%) Total 1990
104 Slovenia	2.0	20	6,540	. .	. .	. .	73	. .	. .
105 Puerto Rico	3.6	9	6,590	0.9	6.5	3.3	74	. .	. .
106 Korea, Rep.	43.7	99	6,790	8.5	20.1	5.9	71	7	4
107 Greece	10.3	132	7,290	1.0	14.5	17.7	77	11	7
108 Portugal	9.8	92	7,450	3.1	16.7	17.4	74	19	15
109 Saudi Arabia	16.8	2,150	7,510	−3.3	24.9	−1.9	69	52	38
Low- and middle-income	4,610.1	101,669	1,040	0.9	26.2	75.7	64	46	36
Sub-Saharan Africa	543.0	24,274	530	−0.8	13.6	15.6	52	62	50
East Asia and Pacific	1,688.8	16,368	760	6.1	16.6	6.7	68	34	24
South Asia	1,177.9	5,133	310	3.0	9.7	8.5	60	69	55
Europe and Central Asia	494.5	24,370	2,080	. .	18.7	47.5	70	. .	. .
Middle East and N. Africa	252.6	11,015	1,950	−2.3	17.0	10.1	64	57	45
Latin America and Caribbean	453.2	20,507	2,690	−0.2	46.7	229.5	68	18	15
Severely indebted	504.6	22,483	2,470	−1.0	42.1	208.0	67	28	23
High-income economies	828.1	31,709	22,160	2.3	9.1	4.3	77	. .	. .
110 Ireland	3.5	70	12,210	3.4	14.2	5.3	75	. .	. .
111 New Zealand	3.4	271	12,300	0.6	12.5	9.4	76	[e]	[e]
112 †Israel	5.1	21	13,220	1.9	39.6	78.9	76	. .	. .
113 Spain	39.1	505	13,970	2.9	16.1	8.7	77	7	5
114 †Hong Kong	5.8	1	15,360[f]	5.5	9.2	7.8	78	. .	. .
115 †Singapore	2.8	1	15,730	5.3	5.9	2.0	75	. .	. .
116 Australia	17.5	7,713	17,260	1.6	11.8	6.4	77	[e]	[e]
117 United Kingdom	57.8	245	17,790	2.4	14.5	5.7	76	[e]	[e]
118 Italy	57.8	301	20,460	2.2	15.6	9.1	77	[e]	[e]
119 Netherlands	15.2	37	20,480	1.7	7.9	1.7	77	[e]	[e]
120 Canada	27.4	9,976	20,710	1.8	8.7	4.1	78	[e]	[e]
121 Belgium	10.0	31	20,880	2.0	7.8	4.1	76	[e]	[e]
122 Finland	5.0	338	21,970	2.0	12.3	6.0	75	[e]	[e]
123 †United Arab Emirates	1.7	84	22,020	−4.3	. .	0.8	72	[e]	[e]
124 France	57.4	552	22,260	1.7	10.2	5.4	77	[e]	[e]

	Population (millions) mid-1992	Area (thousands of sq. km)	GNP per Capita[a]		Avg. Annual Rate of Inflation (%)		Life Expectancy at Birth (years) 1992	Adult Illiteracy (%)	
			Dollars 1992	Avg. Ann. Growth (%), 1980–92	1970–80	1980–92		Female 1990	Total 1990
125 Austria	7.9	84	22,380	2.0	6.5	3.6	77	[e]	[e]
126 Germany	80.6	357	23,030	2.4[g]	5.1[g]	2.7[g]	76	[e]	[e]
127 United States	255.4	9,373	23,240	1.7	7.5	3.9	77	[e]	[e]
128 Norway	4.3	324	25,820	2.2	8.4	4.9	77	[e]	[e]
129 Denmark	5.2	43	26,000	2.1	10.1	4.9	75	[e]	[e]
130 Sweden	8.7	450	27,010	1.5	10.0	7.2	78	[e]	[e]
131 Japan	124.5	378	28,190	3.6	8.5	1.5	79	[e]	[e]
132 Switzerland	6.9	41	36,080	1.4	5.0	3.8	78	[e]	[e]
World	5,438.2	133,378	4,280	1.2	11.6	17.2	66	45	35

†Economies classified by the United Nations or otherwise regarded by their authorities as developing.
[a]In all tables GDP and GNP data cover mainland Tanzania only.
[b]Estimates for economies of the former Soviet Union are subject to more than usual range of uncertainty and should be regarded as very preliminary.
[c]Data reflect recent revision of 1992 GNP per capita: from $700 to $670 for Côte d'Ivoire, from $2,510 to $2,730 for Chile, and from $2,700 to $2,670 for South Africa.
[d]In all tables, data for Jordan cover the East Bank only.
[e]According to UNESCO, illiteracy is less than 5 percent.
[f]Data refer to GDP.
[g]Data refer to the Federal Republic of Germany before unification.

APPENDIX XVIII

SOURCES OF ENVIRONMENTAL AND NATURAL RESOURCE INFORMATION FOR VARIOUS COUNTRIES

	INFOTERRA Member	*Sources of National Environmental Information (a)*							
		National State of the Environment Report	UNCED National Reports	Country Environmental Profile	Environmental Synopses	Biological Diversity Profile	National Conservation Strategy	Environmental Action Plan	Tropical Forestry Action Plan
AFRICA									
Algeria	Yes		1991						
Angola	Yes		1991		IP				
Benin	Yes		1991		IP			IP	
Botswana	Yes	1990	1992			1991	1990		
Burkina Faso	Yes		1991	1980, 1982	IP			IP	FSR IP
Burundi	Yes		1991	1981	IP			1991	
Cameroon	Yes		1992	1981					TFAP 1988, 1989
Central African Rep	Yes		1991		IP				FSR IP
Chad	Yes		1991				1990		
Congo	Yes		1992		IP			IP	FSR IP
Cote d'Ivoire	Yes	1992	1991		IP	1991		IP	FSR IP
Djibouti	No		1991		IP				
Egypt	Yes		1991	1980					
Equatorial Guinea	No		1991		IP				FSR IP
Ethiopia	Yes	IP	1992		IP	1991	1990		FSR IP
Gabon	Yes		1991		IP			IP	FSR IP
Gambia, The	Yes		1991	1981				IP	FSR IP
Ghana	Yes		1991	1980	1992	1988		1991	FSR 1988
Guinea	Yes		1991	1983		1988		IP	TFAP 1988
Guinea-Bissau	Yes		1991			1991	IP	IP	FSR IP
Kenya	Yes	1987	1991			1988	IP		FSR 1987, TFAP IP
Lesotho	Yes		1992	1982	IP			1989	FSR IP
Liberia	Yes			1980	IP				
Libya	Yes								
Madagascar	Yes	1987				1991	1984	1988	FSR IP

Sources of National Environmental Information (a)

	INFOTERRA Member	National State of the Environment Report	UNCED National Reports	Country Environmental Profile	Environmental Synopses	Biological Diversity Profile	National Conservation Strategy	Environmental Action Plan	Tropical Forestry Action Plan
Malawi	Yes		1992	1982	IP		IP		
Mali	Yes		1991	1980, 1989					FSR IP
Mauritania	Yes		1991	1979, 1981	IP		1987		FSR IP
Mauritius	Yes		1991		IP			1990	
Morocco	Yes	IP	1992	1980					
Mozambique	Yes		1991		IP				
Namibia	No	1991	IP		IP				
Niger	Yes		1991	1980	IP				FSR IP
Nigeria	Yes	1991	1991		IP	1988	1986	IP	FSR IP
Rwanda	Yes		1991	1981, 1987	IP			1991	FSR IP
Senegal	Yes		1991	1980		1991			FSR IP
Sierra Leone	No		1991		IP		1985		TFAP 1990
Somalia	Yes			1979	IP		1990		FSR 1990
South Africa	No	1989	1991				1980		
Sudan	Yes		1991	1989					FSR 1986
Swaziland	No		1991	1980	IP				
Tanzania	Yes		1991			1988	1986		TFAP 1989
Togo	Yes		1992		IP		1985	IP	FSR IP
Tunisia	Yes		1992	1980					
Uganda	Yes		1991	1982	1992	1988	1983	IP	
Zaire	Yes		1991	1981	IP	1988	1984		TFAP 1988, 1990
Zambia	Yes	1988	1992	1982			1985		FSR IP
Zimbabwe	Yes	1988	1991	1982	1992		1987		FSR IP

	INFOTERRA Member	Sources of National Environmental Information (a)							
		National State of the Environment Report	UNCED National Reports	Country Environmental Profile	Environmental Synopses	Biological Diversity Profile	National Conservation Strategy	Environmental Action Plan	Tropical Forestry Action Plan
ASIA									
Afghanistan	No		1992						
Bangladesh	Yes	1987	1991	1980			1987, 1991	IP	MPFD IP
Bhutan	Yes		1991				IP	IP	MPFD IP
Cambodia	No			1989					
China	Yes		1991				1990	IP	
India	Yes	1988, 1991, 1991	1992	1980		1989	IP	IP	TFAP IP
Indonesia	Yes	1991	1991	1987	IP		IP	IP	TFAP 1991
Iran, Islamic Rep	Yes		1992						
Iraq	Yes		1992						
Israel	Yes	1979, 1988, 1992	1992						
Japan	Yes	1991	1991						
Jordan	Yes	1989	1992	1979			IP		
Korea, Dem People's Rep	Yes		1991						
Korea, Rep	Yes		1991						
Kuwait	Yes								
Lao People's Dem Rep	No		1992						TFAP 1990
Lebanon	Yes		1991						
Malaysia	Yes	1990, 1991	1992		IP	1988 (2 docs)	1988, 1989, 1991		TFAP 1992
Mongolia	Yes		1991						
Myanmar	No		1992	1982		1989			MPFD 1991
Nepal	Yes	IP	1992	1979			1987	IP	MPFD 1988
Oman	Yes	IP	1992	1981			IP		
Pakistan	Yes	1986	1991	1986, 1988	1992	1991	1991		MPFD IP
Philippines	Yes		1991	1980		1988	IP	1988	MPFD 1990
Saudi Arabia	Yes		1992						

Sources of National Environmental Information (*a*)

	INFOTERRA Member	National State of the Environment Report	UNCED National Reports	Country Environmental Profile	Environmental Synopses	Biological Diversity Profile	National Conservation Strategy	Environmental Action Plan	Tropical Forestry Action Plan
Singapore	Yes	1989, 1990	1991	1988			1990		
Sri Lanka	Yes	1989	1991	1988	1992		1988	1991	MPFD 1986
Syrian Arab Rep	Yes	IP	1992	1981					
Thailand	Yes	IP	1992	1987			IP		MPFD IP
Turkey	Yes	1989	1991						
United Arab Emirates	Yes								
Viet Nam	Yes		1991				1985	IP	TFAP 1991
Yemen	Yes		1991	1982					

World Resources 1994–1995.

	INFOTERRA Member	National State of the Environment Report	UNCED National Reports	Country Environmental Profile	Environmental Synopses	Biological Diversity Profile	National Conservation Strategy	Environmental Action Plan	Tropical Forestry Action Plan
					Sources of National Environmental Information (a)				
NORTH & CENTRAL AMERICA									
Belize	Yes		1992	1984	IP		IP		TFAP 1989
Canada	Yes	1986, IP	1991				1986		
Costa Rica	Yes	1988	1991	1982			1990		TFAP 1990
Cuba	Yes		1991						TFAP 1991
Dominican Rep	No		1991	1981	IP				TFAP 1990
El Salvador	Yes		1992	1985				IP	
Guatemala	Yes		1992	1984			IP		TFAP 1991
Haiti	Yes	IP	1992	1985				IP	FSR IP
Honduras	Yes		1991	1982, 1989				IP	TFAP 1988
Jamaica	Yes	IP	1992	1987	IP				TFAP 1990
Mexico	Yes	1986–1990 (7 docs)	1992			1988			TFAP 1991
Nicaragua	No		1992	1981			IP	IP	FSR IP
Panama	Yes	1985	1991	1980					TFAP 1990
Trinidad and Tobago	No		1992				IP		FSR IP
United States	Yes	1989, 1990, 1991	1992						
SOUTH AMERICA									
Argentina	Yes	IP	1991						TFAP 1988
Bolivia	Yes		1992	1986	IP			IP	TFAP 1989
Brazil	Yes		1991			1988			
Chile	Yes	1990	1992	1990					FSR IP
Colombia	Yes		1992	1990	IP	1988	IP		TFAP 1989
Ecuador	Yes	IP	1991	1987		1988	IP	IP	TFAP 1990
Guyana	Yes		1992	1982					TFAP 1989
Paraguay	Yes		1991	1985				IP	
Peru	Yes		1992	1986	IP	1988	IP		TFAP 1988
Suriname			1992						FSR IP
Uruguay	Yes	1992	1991						
Venezuela	Yes	1990, 1992	1992						FSR IP

	INFOTERRA Member	*Sources of National Environmental Information (a)* National State of the Environment Report	UNCED National Reports	Country Environmental Profile	Environmental Synopses	Biological Diversity Profile	National Conservation Strategy	Environmental Action Plan	Tropical Forestry Action Plan
EUROPE									
Austria	Yes	1988, 1989, 1991	1992						
Belgium	Yes	1979	1992						
Bulgaria	Yes		1992						
Czechoslovakia (former)	Yes	1990[b]	1991						
Denmark	Yes	1982	1991						
Finland	Yes	1988, IP	1992						
France	Yes	1990, 1993	1991				IP		
Germany	Yes	1989, 1990	1992						
Greece	Yes	1983	1991						
Hungary	Yes	1990	1991						
Iceland	Yes	1986	1992						
Ireland	Yes	1985	1992						
Italy	Yes	1989	1991				ND		
Netherlands	Yes	1989, 1990	1991						
Norway	Yes	1993	1992				IP		
Poland	Yes	1989, 1990	1991						
Portugal	Yes	1989	1991						
Romania	Yes		1992						
Spain	Yes	1977	1991				IP		
Sweden	Yes	1984, 1990, 1993	1992						
Switzerland	Yes	1989	1992				IP		
United Kingdom	Yes	1990, 1992	1992				1983, 1990		
Yugoslavia (former)	Yes	1987	1991[c]				IP		

		Sources of National Environmental Information (a)							
	INFOTERRA Member	National State of the Environment Report	UNCED National Reports	Country Environmental Profile	Environmental Synopses	Biological Diversity Profile	National Conservation Strategy	Environmental Action Plan	Tropical Forestry Action Plan
U.S.S.R. (FORMER)	Yes	1988	1991						
Armenia									
Azerbaijan			1992						
Belarus			1992						
Estonia			1992						
Georgia									
Kazakhstan			1992						
Kyrgyzstan									
Latvia			1992						
Lithuania			1992						
Moldova									
Russian Federation			1992						
Tajikistan									
Turkmenistan									
Ukraine			1991						
Uzbekistan									
OCEANIA									
Australia	Yes	1987, 1988, 1992	1991				1988		
Fiji	Yes		1992		IP	1989	IP		FSR 1990
New Zealand	Yes	1988	1991				1985		
Papua New Guinea	Yes		1992		IP				TFAP 1990
Solomon Islands	No		1991		IP				

Sources: World Resources Institute, International Institute for Environment and Development, World Conservation Union, U.S. Agency for International Development, World Conservation Monitoring Centre, and the United Nations Environment Programme.

Notes: a. Publication date of most recent edition; multiple dates indicate different reports. b. Produced by and about the Czech Republic. c. Yugoslavia 1991 and Croatia 1992. INFOTERRA: member of INFOTERRA, the global environmental information system; FSR = Forestry Sector Review; TFAP = Tropical Forestry Action Plan; MPFD = Master Plan for Forestry Development; IP = in preparation; ND = published, no date.

APPENDIX XIX

SOVEREIGN NATIONS' COOPERATION AND PARTICIPATION IN MAJOR GLOBAL CONVENTIONS

	Global Conventions								
	ATMOSPHERE			HAZARDOUS SUBSTANCES				REGIONAL AGREEMENTS (b)	
	Ozone Layer 1985	CFC Control 1987	Climate Change (a) 1992	Biological and Toxin Weapons 1972	Nuclear Accident Notification 1986	Nuclear Accident Assistance 1986	Hazardous Waste Movement (a) 1989	UNEP Regional Seas	Other Regional Agreements
WORLD									
AFRICA									
Algeria	CP	CP	CP		S	S		M	AFC
Angola			S						
Benin	CP	CP	S	CP				WCA	AFC, HW
Botswana	CP	CP	S	CP					AFC
Burkina Faso	CP	CP	CP	CP					HW
Burundi			S	S					AFC, HW
Cameroon	CP	CP	S		S	S		WCA	AFC, HW
Central African Rep	CP		S	S					AFC, HW
Chad	CP		S						AFC, HW
Congo		S	S	CP				WCA	AFC
Cote d'Ivoire	CP	CP	S	S	S	S		WCA	AFC, HW
Djibouti			S						AFC, HW
Egypt	CP	CP	S	S	CP	CP		M, RS	AFC, HW
Equatorial Guinea	CP			CP					
Ethiopia			S	CP					AFC
Gabon			S	S				WCA	AFC
Gambia, The	CP	CP	S	S				WCA	AFC
Ghana	CP	CP	S	CP				WCA	AFC
Guinea	CP	CP	CP					WCA	AFC, HW
Guinea-Bissau			S	CP					HW
Kenya	CP	CP	S	CP				EA	AFC
Lesotho			S	CP					AFC, HW
Liberia			S					WCA	AFC
Libya	CP	CP	S	CP		CP		M	AFC, HW
Madagascar			S	S				EA	AFC

	Global Conventions								
	ATMOSPHERE			HAZARDOUS SUBSTANCES				REGIONAL AGREEMENTS (b)	
	Ozone Layer 1985	CFC Control 1987	Climate Change (a) 1992	Biological and Toxin Weapons 1972	Nuclear Accident Notification 1986	Nuclear Accident Assistance 1986	Hazardous Waste Movement (a) 1989	UNEP Regional Seas	Other Regional Agreements
Malawi	CP	CP	S	S					AFC
Mali			S	S	S	S			AFC, HW
Mauritania			S					WCA	AFC
Mauritius			S	CP					AFC
Morocco	S	S	S	S	S	S		M	AFC
Mozambique			S						AFC
Namibia			S						
Niger	CP		S	CP	S	S			AFC, HW
Nigeria	CP	CP	S	CP	CP	CP	CP	WCA	AFC
Rwanda			S	CP					AFC, HW
Senegal		CP	S	CP	S	S	CP	WCA	AFC, EC, HW
Sierra Leone			S	CP	S	S			AFC
Somalia				S				EA, RS	AFC, HW
South Africa	CP	CP	S	CP	CP	CP			
Sudan	CP	CP	S		S	S		RS	AFC
Swaziland	CP	CP	S						AFC, HW
Tanzania	CP	CP	S	S			CP		AFC, HW
Togo	CP	CP	S	CP				WCA	AFC, HW
Tunisia	CP	CP	CP	CP	CP	CP		M	AFC, HW
Uganda	CP	CP	CP						AFC
Zaire			S	CP	S	S			AFC
Zambia	CP	CP	CP						AFC
Zimbabwe	CP	CP	CP	CP	S	S			HW

	Global Conventions								
	ATMOSPHERE			HAZARDOUS SUBSTANCES				REGIONAL AGREEMENTS (b)	
	Ozone Layer 1985	CFC Control 1987	Climate Change (a) 1992	Biological and Toxin Weapons 1972	Nuclear Accident Notification 1986	Nuclear Accident Assistance 1986	Hazardous Waste Movement (a) 1989	UNEP Regional Seas	Other Regional Agreements
ASIA									
Afghanistan			S	CP	S	S	S		
Bangladesh	CP	CP	S	CP	CP	CP	CP		
Bhutan			S	CP					
Cambodia				CP					
China	CP	CP	CP	CP	CP	CP	CP		
India	CP	CP	CP	CP	CP	CP	CP		
Indonesia	CP	CP	S	CP	S	S			ASC
Iran, Islamic Rep	CP	CP	S	CP	S	S	CP	K	
Iraq				CP	CP	CP		K	
Israel	CP	CP	S		CP	CP	S	M	
Japan	CP	CP	CP	CP	CP	CP			
Jordan	CP	CP	CP	CP	CP	CP	CP	RS	
Korea, Dem People's Rep			S	CP	S	S			
Korea, Rep	CP		S	CP	CP	CP			
Kuwait	CP			CP			S	K	
Lao People's Dem Rep				CP					
Lebanon	CP		S	CP	S	S	S	M	
Malaysia	CP	CP	S	CP	CP	CP			ACS
Mongolia			CP	CP	CP	CP			
Myanmar			S	S					
Nepal			S	S					
Oman			S	CP				K	
Pakistan	CP	CP	S	CP		CP			
Philippines	CP	CP	S	CP			S		ASC
Saudi Arabia	CP	CP		CP	CP	CP	CP	K, RS	

	Global Conventions								
	ATMOSPHERE			HAZARDOUS SUBSTANCES				REGIONAL AGREEMENTS (b)	
	Ozone Layer 1985	CFC Control 1987	Climate Change (a) 1992	Biological and Toxin Weapons 1972	Nuclear Accident Notification 1986	Nuclear Accident Assistance 1986	Hazardous Waste Movement (a) 1989	UNEP Regional Seas	Other Regional Agreements
Singapore	CP	CP	S	CP					ASC
Sri Lanka	CP	CP	S	CP	CP	CP	CP		
Syrian Arab Rep	CP	CP		S	S	S	CP	M	
Thailand	CP	CP	S	CP	CP	CP	S		ASC
Turkey	CP	CP		CP	S	CP	S	M, BS	EC, LR
United Arab Emirates	CP	CP		S	CP	CP	S	K	
Viet Nam			S	CP	CP	CP			
Yemen			S					RS	
NORTH AND CENTRAL AMERICA									
Belize			S	CP					
Canada	CP	CP	CP	CP	CP	S	CP		LR, EIA
Costa Rica	CP	CP	S	CP	CP	S			
Cuba	CP	CP	S	CP	S	CP		C	
Dominican Rep	CP	CP	S	CP					
El Salvador	CP		S	CP			CP		
Guatemala	CP	CP	S	CP	CP	CP	S	C	
Haiti			S	S			S		
Honduras			S	CP				C	
Jamaica	CP	CP	S	CP				C	
Mexico	CP	CP	CP	CP	CP	CP	CP	C	
Nicaragua	CP	CP	S	CP				C	
Panama	CP	CP	S	CP	S	S	CP	SEP, C	
Trinidad and Tobago	CP	CP	S					C	
United States	CP	CP	CP	CP	CP	CP	S	C, SP	LR, EIA

	Global Conventions								
	ATMOSPHERE			HAZARDOUS SUBSTANCES				REGIONAL AGREEMENTS (b)	
	Ozone Layer 1985	CFC Control 1987	Climate Change (a) 1992	Biological and Toxin Weapons 1972	Nuclear Accident Notification 1986	Nuclear Accident Assistance 1986	Hazardous Waste Movement (a) 1989	UNEP Regional Seas	Other Regional Agreements
SOUTH AMERICA									
Argentina	CP	CP	S	CP	CP	CP	CP		
Bolivia			S	CP			S		AMC
Brazil	CP	CP	S	CP	CP	CP	CP		AMC
Chile	CP	CP	S	CP	S	S	CP	SEP	
Colombia	CP		S				S	SEP, C	AMC
Ecuador	CP	CP	CP	CP			CP	SEP	AMC
Guyana			S	S					AMC
Paraguay	CP	CP	S	CP	S	S			
Peru	CP	CP	CP	CP				SEP	AMC
Suriname			S	CP					AMC
Uruguay	CP	CP	S	CP	CP	CP	CP		
Venezuela	CP	CP	S	CP			S	C	AMC
EUROPE									
Albania				CP				M	EIA
Austria	CP	CP	S	CP	CP	CP	CP		EC, LR, EIA
Belgium	CP	CP	S	CP	S	S	S		EC, LR, EIA
Bulgaria	CP	CP	S	CP	CP	CP		BS	EC, LR, EIA
Czechoslovakia (former)			CP c		CP d	CP d			LR, EIA
Denmark	CP	CP	S	CP	CP	S	S		EC, LR, EIA
Finland	CP	CP	S	CP	CP	CP	CP		EC, LR, EIA
France	CP	CP	S	CP	CP	CP	CP	M, C, EA, SP	EC, SPC, LR, EIA
Germany	CP	CP	S	CP	CP	CP	S		EC, LR, EIA
Greece	CP	CP	S	CP	CP	CP	S	M	EC, R, EIA

	Global Conventions								
	ATMOSPHERE			HAZARDOUS SUBSTANCES				REGIONAL AGREEMENTS (b)	
	Ozone Layer 1985	CFC Control 1987	Climate Change (a) 1992	Biological and Toxin Weapons 1972	Nuclear Accident Notification 1986	Nuclear Accident Assistance 1986	Hazardous Waste Movement (a) 1989	UNEP Regional Seas	Other Regional Agreements
Hungary	CP	CP	S	CP	CP	CP	CP		EC, LR, EIA
Iceland	CP	CP	CP	CP	CP	S			LR, EIA
Ireland	CP	CP	S	CP	CP	CP	S		EC, LR, EIA
Italy	CP	CP	S	CP	CP	CP	S	M	EC, LR, EIA
Netherlands	CP	CP	S	CP	CP	CP	CP	C (e)	EC, LR, EIA
Norway	CP	CP	CP	CP	CP	CP	CP		EC, LR, EIA
Poland	CP	CP	S	CP	CP	CP	CP		LR, EIA
Portugal	CP	CP	S	CP	S	S	S		EC, LR, EIA
Romania	CP	CP	S	CP	CP	CP	CP	BS	LR, EIA
Spain	CP	CP	S	CP	CP	CP	S	M	EC, LR, EIA
Sweden	CP	CP	CP	CP	CP	CP	CP		EC, LR, EIA
Switzerland	CP	CP	S	CP	CP	CP	CP		EC, LR
United Kingdom	CP	CP	S	CP	CP	CP	S	C (f), SP	EC, LR, EIA
Yugoslavia (former)	CP g	CP g	S h	CP i	CP g	CP g		M	LR
U.S.S.R. (former)									
Armenia			CP						
Azerbaijan			S						
Belarus	CP	CP	S	CP	CP	CP			LR, EIA
Estonia			S				CP		
Georgia								BS	
Kazakhstan			S						
Kyrgyzstan									
Latvia			S		CP	CP	CP		
Lithuania			S						
Moldova			S						

	Global Conventions								
	ATMOSPHERE			HAZARDOUS SUBSTANCES				REGIONAL AGREEMENTS (b)	
	Ozone Layer 1985	CFC Control 1987	Climate Change (a) 1992	Biological and Toxin Weapons 1972	Nuclear Accident Notification 1986	Nuclear Accident Assistance 1986	Hazardous Waste Movement (a) 1989	UNEP Regional Seas	Other Regional Agreements
Russian Federation	CP	CP	S	CP	CP	CP	CP	BS	LR, EIA
Tajikistan									
Turkmenistan									
Ukraine	CP	CP	S	CP	CP	CP		BS	LR, EIA
Uzbekistan	CP	CP	CP						
OCEANIA									
Australia	CP	CP	CP	CP	CP	CP	CP	SP	SPC
Fiji	CP	CP	CP	CP				SP	SPC
New Zealand	CP	CP	CP	CP	CP	CP	S	SP	
Papua New Guinea	CP	CP	CP	CP				SP	SPC
Solomon Islands	CP	CP	S	CP				SP	

	WILDLIFE AND HABITAT						OCEANS		
	Antarctic Treaty and Convention 1959 & 1980	Wetlands (Ramsar) 1971	World Heritage 1972	Endangered Species (CITES) 1973	Migratory Species 1979	Biodiversity 1992	Ocean Dumping 1972	Ship Pollution (MARPOL) 1978	Law of the Sea (a) 1982
WORLD									
AFRICA									
Algeria		CP	CP	CP		S		CP	S
Angola			CP			S			CP
Benin			CP	CP	CP	S			S
Botswana				CP		S			CP
Burkina Faso		CP	CP	CP	CP	CP			S
Burundi			CP	CP		S			S
Cameroon			CP	CP	CP	S			CP
Central African Rep			CP	CP	S	S			S
Chad		CP		CP	S	S	S		S
Congo			CP	CP		S			S
Cote d'Ivoire			CP		S	S	CP	CP	CP
Djibouti				CP		S		CP	CP
Egypt		CP	CP	CP	CP	S		CP	CP
Equatorial Guinea									S
Ethiopia			CP	CP		S			S
Gabon		CP	CP	CP		S	CP	CP	S
Gambia, The			CP	CP		S		CP	CP
Ghana		CP	CP	CP	CP	S		CP	CP
Guinea		CP	CP	CP		CP			CP
Guinea-Bissau		CP		CP		S			CP
Kenya		CP	CP	CP		S	CP	CP	CP
Lesotho		CP		S		S	S		S
Liberia				CP		S	S	CP	S
Libya			CP			S	CP		S
Madagascar			CP	CP	S	S			S

	WILDLIFE AND HABITAT					OCEANS		
	Antarctic Treaty and Convention 1959 & 1980	Wetlands (Ramsar) 1971	World Heritage 1972	Endangered Species (CITES) 1973	Migratory Species 1979	Biodiversity 1992	Ocean Dumping 1972	Ship Pollution (MARPOL) 1978	Law of the Sea (a) 1982
Malawi			CP	CP		S			S
Mali		CP	CP		CP	S			CP
Mauritania		CP	CP			S			S
Mauritius				CP		S			S
Morocco		CP	CP	CP	S	S	CP		S
Mozambique			CP	CP		S			S
Namibia				CP		S			CP
Niger		CP	CP	CP	CP	S			S
Nigeria			CP	CP	CP	S	CP		CP
Rwanda				CP		S			S
Senegal		CP	CP	CP	CP	S	S		CP
Sierra Leone									S
Somalia				CP	CP		S		CP
South Africa	CP, MLR	CP		CP	CP	S	CP	CP	S
Sudan			CP	CP		S			CP
Swaziland						S			S
Tanzania			CP	CP		S			CP
Togo				CP	S	S	S	CP	CP
Tunisia		CP	CP	CP	CP	CP	CP	CP	CP
Uganda		CP	CP	CP	S	CP			CP
Zaire			CP	CP	CP	S	CP		CP
Zambia		CP	CP	CP		CP			CP
Zimbabwe			CP	CP		S			S

	WILDLIFE AND HABITAT						OCEANS		
	Antarctic Treaty and Convention 1959 & 1980	Wetlands (Ramsar) 1971	World Heritage 1972	Endangered Species (CITES) 1973	Migratory Species 1979	Biodiversity 1992	Ocean Dumping 1972	Ship Pollution (MARPOL) 1978	Law of the Sea (a) 1982
ASIA									
Afghanistan			CP	CP		S	CP		S
Bangladesh		CP	CP		S			S	
Bhutan						S			S
Cambodia			CP	S			S		S
China	CP	CP	CP	CP		CP	CP	CP	S
India	CP, MLR	CP	CP	CP	CP	CP		CP	S
Indonesia			CP	CP		S		CP	CP
Iran, Islamic Rep		CP	CP	CP		S			S
Iraq			CP			S			CP
Israel				CP	CP			CP	
Japan	CP, MLR	CP	CP		CP	CP	CP	S	
Jordan		CP	CP	CP		CP	CP		
Korea, Dem People's Rep	CP					S		CP	S
Korea, Rep	CP, MLR		CP	CP		S		CP	S
Kuwait				S		S	S		CP
Lao People's Dem Rep			CP						S
Lebanon			CP			S	S	CP	S
Malaysia			CP	CP		S			S
Mongolia			CP			CP			S
Myanmar						S		CP	S
Nepal		CP	CP	CP		S	S		S
Oman			CP			S	CP	CP	CP
Pakistan		CP	CP	CP	CP	S			S
Philippines			CP	CP	S	CP	CP		CP
Singapore						S		CP	

	WILDLIFE AND HABITAT						OCEANS		
	Antarctic Treaty and Convention 1959 & 1980	Wetlands (Ramsar) 1971	World Heritage 1972	Endangered Species (CITES) 1973	Migratory Species 1979	Biodiversity 1992	Ocean Dumping 1972	Ship Pollution (MARPOL) 1978	Law of the Sea (a) 1982
Saudi Arabia			CP		CP				S
Sri Lanka		CP	CP	CP	CP	S			S
Syrian Arab Rep			CP			S		CP	
Thailand			CP	CP		S			S
Turkey			CP			S		CP	
United Arab Emirates				CP		S	CP		S
Viet Nam		CP	CP	S		S		CP	S
Yemen			CP			S			S
NORTH AND CENTRAL AMERICA									
Belize			CP	CP		S			CP
Canada	CP, MLR	CP	CP	CP		CP	CP	CP	S
Costa Rica		CP	CP	CP		S	CP		CP
Cuba	CP		CP	CP		S	CP	CP	CP
Dominican Rep			CP	CP		S	CP		S
El Salvador			CP	CP		S			S
Guatemala	CP	CP	CP	CP		S	CP		S
Haiti			CP			S	CP		S
Honduras			CP	CP		S	CP		S
Jamaica			CP		S	S	CP	CP	CP
Mexico		CP	CP	CP		CP	CP	CP	CP
Nicaragua			CP	CP		S			S
Panama		CP	CP	CP	CP	S	CP	CP	S
Trinidad and Tobago		CP		CP		S			CP
United States	CP, MLR	CP	CP	CP		S	CP	CP	

	WILDLIFE AND HABITAT						OCEANS		
	Antarctic Treaty and Convention 1959 & 1980	Wetlands (Ramsar) 1971	World Heritage 1972	Endangered Species (CITES) 1973	Migratory Species 1979	Biodiversity 1992	Ocean Dumping 1972	Ship Pollution (MARPOL) 1978	Law of the Sea (a) 1982
SOUTH AMERICA									
Argentina	CP, MLR	CP	CP	CP		S	CP		S
Bolivia		CP	CP	CP		S	S		S
Brazil	CP, MLR		CP	CP		S	CP	CP	CP
Chile	CP, MLR	CP	CP	CP	CP	S	CP		S
Colombia	CP		CP	CP		S	S	CP	S
Ecuador	CP	CP	CP	CP		CP		CP	
Guyana			CP	CP		S			S
Paraguay		S	CP	CP		S			CP
Peru	CP, MLR	CP	CP	CP		CP		CP	
Suriname		CP		CP	CP	S	CP	CP	S
Uruguay	CP, MLR	CP	CP	CP	CP	CP	S	CP	CP
Venezuela		CP	CP	CP		S	S		
EUROPE									
Albania			CP						
Austria	CP	CP	CP	CP		S		CP	S
Belgium	CP, MLR	CP		CP	CP	S	CP	CP	S
Bulgaria	CP, MLR	CP	CP	CP		S		CP	S
Czechoslovakia (former)		CP b	CP b	CP b		S b		CP	S
Denmark	CP	CP	CP	CP	CP	S	CP	CP	S
Finland	CP, MLR	CP	CP	CP	CP	S	CP	CP	S
France	CP, MLR	CP	CP	CP	CP	S	CP	CP	S
Germany	CP, MLR	CP	CP	CP	CP	S	CP	CP	
Greece	CP, MLR	CP	CP	CP	S	S	CP	CP	S

	WILDLIFE AND HABITAT						OCEANS		
	Antarctic Treaty and Convention 1959 & 1980	Wetlands (Ramsar) 1971	World Heritage 1972	Endangered Species (CITES) 1973	Migratory Species 1979	Biodiversity 1992	Ocean Dumping 1972	Ship Pollution (MARPOL) 1978	Law of the Sea (a) 1982
Hungary	CP	CP	CP	CP	CP	S	CP	CP	S
Iceland		CP				S	CP	CP	CP
Ireland		CP	CP	S	CP	S	CP		S
Italy	CP, MLR	CP	CP	CP	CP	S	CP	CP	S
Netherlands	CP, MLR	CP	CP	CP	CP	S	CP	CP	S
Norway	CP, MLR	CP	CP	CP	CP	CP	CP	CP	S
Poland	CP, MLR	CP	CP	CP		S	CP	CP	S
Portugal		CP	CP	CP	CP	S	CP	CP	S
Romania	CP	CP	CP			S			S
Spain	CP, MLR	CP	CP	CP	CP	S	CP	CP	S
Sweden	CP, MLR	CP	CP	CP	CP	S	CP	CP	S
Switzerland	CP	CP	CP	CP		S	CP	CP	S
United Kingdom	CP, MLR	CP	CP	CP	CP	S	CP	CP	
Yugoslavia (former)		CP c	CP d			S e	CP c	CP c	CP
U.S.S.R. (former)									
Armenia		CP				CP			
Azerbaijan		CP				S			
Belarus			CP			CP	CP		
Estonia		CP		CP		S		CP	
Georgia		CP	CP						
Kazakhstan		CP				S			
Kyrgyzstan		CP							
Latvia						S			
Lithuania		CP	CP			S		CP	
Moldova		CP				S			

	WILDLIFE AND HABITAT						OCEANS		
	Antarctic Treaty and Convention 1959 & 1980	Wetlands (Ramsar) 1971	World Heritage 1972	Endangered Species (CITES) 1973	Migratory Species 1979	Biodiversity 1992	Ocean Dumping 1972	Ship Pollution (MARPOL) 1978	Law of the Sea (a) 1982
Russian Federation	CP, MLR	CP	CP	CP		S	CP	CP	S
Tajikistan		CP	S						
Turkmenistan						S			
Ukraine			CP			S	CP		
Uzbekistan		CP	CP						
OCEANIA									
Australia	CP, MLR	CP	CP	CP	CP	CP	CP	CP	S
Fiji			CP			CP			CP
New Zealand	CP, MLR	CP	CP	CP		CP	CP		S
Papua New Guinea	CP			CP		CP	CP		S
Solomon Islands			CP			S	CP		S

Sources: Environmental Law Information System of the World Conservation Union Environmental Law Centre and the United Nations Environment Programme (UNEP).

Notes: a. Convention not yet in force. b. Regional agreement letter codes (M, ML, etc.) indicate ratification of specific regional agreements. c. Czech Republic; the Slovak Republic is a signatory. d. Slovak Republic. e. Ratified on behalf of Aruba and the Netherlands Antilles Federation. f. Ratified on behalf of the British Virgin Islands. Cayman Islands, and the Turks and Caicos Islands. g. Croatia, Slovenia, and Yugoslavia. h. The constituent republics of the former Yugoslavia inherited the status of signatories. i. Slovenia and Yugoslavia. CP = contracting party (has ratified or taken equivalent action); S = signatory; + = has signed or ratified at least two protocols to this convention, * = signatory to regional agreement. UNEP Regional Seas agreements: BS = Black Sea convention; M = Mediterranean convention; WCA = West and Central African convention; EA = East African convention; RS = Red Sea and Gulf of Aden convention; C = Caribbean convention; SEP = South-East Pacific convention; SP = South Pacific convention; K = Kuwait convention. Other Regional Agreements: AFC = African conservation convention; HW = African hazardous waste convention; EC = European conservation convention; LR = transboundary air pollution convention; EIA = environmental impact assessment convention; AMC = Amazonian cooperation treaty; ASC = ASEAN conservation agreement; SPC = South Pacific conservation convention. Some small countries are not included in this table.

Source: World Resources 1994–1995, A Guide to Global Environment, World Resource Institute 1995.

BIBLIOGRAPHY

JOURNAL ARTICLES

Allen, Arletha Vickers. "Consultants Clean Up." *Black Enterprise* 24, No. 5 (December 1993): 43.

"Beware of Rosy Forecasts, Say Environmental Experts." *Civil Engineering* 63, No. 12 (December 1993): 19–20.

Brankert, George G. "The Environment, the Moralist, the Corporation and Its Culture." *Business Ethics Quarterly* 5, No. 4 (October 1995): 675–697.

Brinkley, Douglas. "Bringing the Green Revolution to Africa." *World Policy Journal* 13, No. 1 (spring 1996): 53–62.

Carlson, D. A., and A. M. Sholtz. "Lessons from Southern California for Environmental Markets." *Environmental Law Practice* 1, No. 4, (1994): 15–26.

"Companies to Watch: Developing in a Developing Nation." *Datamation* 34, No. 9 (May 1, 1988): 70.

Daly, H., and R. Goodland. "An Ecological-Economic Assessment of Deregulation of International Commerce Under GATT." *Population Environmental Journal* 15, No. 5 (1994): 395–427.

Dodds, Peter J. "The Evolution of an Environmental Monitor." *Civil Engineering* 62, No. 6 (June 1992): 56–58.

Etter, Irvin B. "Good Advice Doesn't Come Cheap." *Safety and Health* 148, No. 2 (August 1993): 3.

Finke, James. "The Making of a Consultant." *Home-Office Computing* 7, No. 1 (January 1989): 46.

Goodrodge, George. "Attention to Contracts Can Reduce the Liability of Environmental Consultants." *Journal of Environmental Health* 54, No. 4 (January 1992): 23.

Hemphill, Thomas A. "The Ungreen Corporation and People of Color: Unequal Protection: Environmental Justice and Communities of Color." Edited by Robert D. Bullard in *Business and Society Review* 91 (fall 1994): 74–77.

Holden, C. "Science Career Trends for the 90s." *Science* (May 24, 1991): 1119.

"Italian Veneer Producer Seeks Cleaner Environment." *Wood Technology* 121, No. 3 (May 1994): 31–32.

Jacobs, Barbara: "The Total Quality Corporation: How 10 Major Companies Turned Quality and Environmental Challenges to Competitive Advantage in the 1990's by Francis McInerney and Sean White" in *Booklist;* October 1, 1995, Vol. 92, No. 3, p. 239.

Kimmerling, George. "How to Start a Consulting Business." *Training and Development* 49, No. 6 (June 1995): 22–28.

Lantos, Peter R. "The Practice of Consulting." *Chemtech* 21, No. 12 (December 1991): 715–717.

Lentini, Fern. "Building a Consulting Practice." *Journal of Accountancy* 172, No. 1 (July 1991): 69–72.

Lindberg, Per. "Future Gas Market in Europe: Opportunity or Dilemma to the Producer." *Proc. Oil Gas Econ Finance Management Conf.* London, England: Society of Petroleum Engineers, 1992, 121–134.

London, Anne. "Discovering Social Issues: Organizational Development in a Multicultural Community." *Journal of Applied Behavioral Science* 28, No. 3 (September 1992): 445–460.

Marsh, Boyd T. "When to Hire a Consultant: A Public Agency Viewpoint." *Journal of Environmental Health* 51, No. 4 (March 1989): 231.

Marsh, Freddie. "Business: When in Tokyo." *World Press Review* 36, No. 5 (May 1989): 55.

McIlvains, R. W. "Air and Waste Management Markets in Shrinking World." *Journal of Air and Waste Management Association* 39 No. 3, 277.

Raul, A. C., Hagen. "The Convergence of Trade and Environmental Law," *Natural Resource and Environment Journal;* 1993, vol. 8, no. 2, pp. 3–6.

Razdan, Maharaj K. "Beware Foreign Winds Blowing over the Airwaves." *IPI Report* 26, No. 10 (July 1995): 64–65.

"Regulations Spur Automated Hazard Management." *Civil Engineering* 62, No. 2 (February 1992): 18–19.

Sinyak, Y. "Global Climate and Energy Systems." *Pollution Abstracts,* Section P9000 Environmental Action, Chemical Societies on Environmental Issues (EURO-ENVIRONMENT '92) (10–14 May 1992), Budapest, Hungary.

Stark, Phyllis. "Pollack Broadens Consulting Borders." *Billboard* 105, No. 3 (January 1993): 67–69.

Teichman, Ron. "The Unfortunately Secret Field of Occupational and Environmental Medicine." *Journal of Community Health* 18, No. 6 (December 1993): 323–326.

Teskey, R. H. *Joint Venturing in Canada.* Edmonton, Alberta Law Firm of Field & Field International Joint Ventures Journal 1990, Canada.

Verespej, Michael A. "Help Wanted." *Industry Week* 243, No. 22 (December 5, 1994): 55.

Wang, Penelope. "Consulting Wars." *Forbes* 142, No. 11 (November 14, 1988): 355.

Weston, Sandra M. C. "Partnering for Environmental Restoration, The Port Hope Harbour Remedial Action Plan," *Environmental Engineering Journal* 1293 (1993): 297–305.

Westra, Laura. "The Corporation and the Environment." *Business Ethics Quarterly* 5, No. 4 (October 1995): 661–673.

White, Louis P. "Ethical Dilemmas in Organization Development: A Cross-Cultural Analysis." *Journal of Business Ethics* 11, No. 9 (September 1992): 663–670.

Williams, Bob. "Caveat Emptor Enviroconsultants." *Oil and Gas Journal* 91, 9 (August 9, 1993): 11.

BOOKS

Argyris, Chris, *Integrating the Individual and the Organization.* New York: John Wiley & Sons, 1964.

Basta, Nicholas. *Environmental Jobs for Scientists and Engineers.* New York: John Wiley & Sons, 1992.

Bellman, Geoffrey M. *The Consultant Calling: Bringing Who You Are to What You Do.* San Francisco: Jossey-Bass, 1990.

Bennis, Warren. *The Unconscious Conspiracy: Why Leaders Can't Lead.* New York: AMACOM, 1975.

Bermont, Herbert Ingram. *The Complete Consultant: A Roadmap to Success.* Washington, D.C.: Consultant's Library, 1982.

Blake, Robert Rogers. *Consultation: A Handbook for Individuals and Organization Development.* 2nd ed. Reading, MA: Addison-Wesley, 1983.

Carter, John D., Robert F. Cushman, and Scott C. Hartz, eds. *The Handbook of Joint Venturing.* Homewood, Ill.: Dow Jones-Irwin, 1988.

Chemical Scientist: Supply and Demand in Changing World. Prepared by Kline & Co., Marketing Division of American Chemical Society, 1990. Washington, D.C.

Cohen, Susan. *Green at Work.* Island Press, Washington, D.C.: 1995.

Cohen, William A. *How to Make It Big as a Consultant.* New York: American Management Association, 1985.

Cohn, Susan. *Green at Work: Finding a Business Career That Works for the Environment.* Washington, D.C.: Island Press, 1992.

Earth Journal: Environmental Almanac and Resource Directory 1993. Boulder, Colorado, Buzzworm Books, 1993.

The Environmental Management Source Book 1993–94. Arlington Heights, IL: Environment Today, 1993.

Feldman, Samuel. *The Big Book of Business Information—Significant Answers from Owner-Managers of Small-Retailing Service, Manufacturing, and Home Business.* New York, NY, Gallery Books, 1987.

Fiedler, Frederick. *A Theory of Leadership Effectiveness.* New York: McGraw Hill, 1967.

Gelder, Alice A. "How to Do Business with 192 Countries by Phone, Fax and Mail." In *World Business Desk Reference.* New York: Richard D. Irwin, 1994.

Graham, John W., and C. H. Wendy. *A Guide to the Corporate and Non-Profit Sector;* New York, NY, 1994.

Greenfield, W. N. *Successful Management Consulting: Building a Practice with Smaller Company Clients.* Englewood Cliffs, N.J.: Prentice-Hall, 1987.

Greiner, Larry E., and Robert O. Metzger. *Consulting to Management.* Englewood Cliffs, N.J.: Prentice Hall, 1983.

Herzberg, Fred. *Work and the Nature of Man.* Cleveland: World, 1966.

Holtz, Herman. *How to Succeed as an Independent Consultant.* New York: John Wiley & Sons, 1983.

Kave, Harvey. *Inside the Technical Consulting Business: Launching and Building Your Independent Practice.* 2nd ed. New York: John Wiley & Sons, 1994.

Kelly, Robert E. *Consulting: The Complete Guide to a Profitable Career.* New York: Scribner, 1981.

Lamprecht, James L. *ISO 9000: Preparing for Registration.* New York: Marcel Dekker, 1995.

Lant, Jeffrey. *The Unabashed Self-Promoter's Guide,* 2nd Ed. Cambridge, MA. JLA Publications, 1992.

Lavin, Michael R. *Business Information—How to Use It, How to Find It.* Phoenix, Arizona: Oryx, 1992.

Maslow, Abraham. *Motivation and Personality.* New York: Harper & Row, 1970.

Moore, Gerald L. *The Politics of Management Consulting.* New York: Praeger, 1984.

Rensis, Likert. *New Patterns of Management.* New York: McGraw Hill, 1961.

Schneider, Jim. *The Feel of Success in Selling.* Englewood Cliffs, NJ: Prentice Hall, 1990.

Schrello, Don M. *The Complete Marketing Handbook for Consultants.* San Diego: University Associates, 1990.

Serageldin, Ismail. *Development Partners: Aid and Cooperation in the 1990s.* SIDA, 1993. Oxford University Press Inc, New York, NY.

Shenson, Howard L. *Shenson on Consulting: Success Strategies from the "Consultant's Consultant,"* New York: John Wiley & Sons, 1990.

Shenson, Howard L. *How to Strategically Negotiate the Consulting Contract.* Washington, D.C.: Bermont Books, 1980.

Sno, C. P. *The Cultures and a Second Look.* Cambridge, England: Cambridge University Press, 1980, 32.

Stryker, Steven C. *Guide to Successful Consulting: With Forms, Letters and Checklists.* Englewood Cliffs, N.J.: Prentice-Hall, 1984.

Stryker, Steven C. *Principles and Practices of Professional Consulting.* Washington, D.C.: Consultant's Library, 1982.

Sunar, D. G. *Getting Started as a Consulting Engineer.* San Carlos, CA, Professional Publications, 1986.

Tregoe, Benjamin B., and John W. Zimmerman. *Top Management Strategy: What It Is and How to Make It Work.* New York: Simon and Schuster, 1980.

Walter, Ingo, and Tracy Murray, eds. *A Handbook of International Management.* New York: John Wiley & Sons, 1990.

Warner, David J. *Environmental Careers: A Practical Guide to Opportunities in the 90s.* Boca Raton, Fla.: Lewis Publishers, 1992.

Weiss, Alan. *Million Dollar Consulting: The Professional's Guide to Growing a Practice.* New York: McGraw Hill, 1992.

World Bank. *World Development Report 1992: Development and the Environment.* New York, NY, Oxford Press, 1992.

The World Bank and the Environment Fiscal 1993, Washington D.C.: 1993.

World Development Report 1994: Infrastructure for Development, Oxford University Press Inc., New York, NY.

The World Directory of Environmental Organizations. 4th ed. California Institute of Public Affairs in Cooperation with Sierra Club and IUCN—The World Conservation Union, 1992. Sacramento, CA.

World Resources Institute. *A Guide to the Global Environment: People and the Environment 1994–1995.* New York, NY, Oxford Press, 1994.

INDEX